DETECTION AT STATE BORDERS
OF NUCLEAR AND OTHER
RADIOACTIVE MATERIAL
OUT OF REGULATORY CONTROL

IAEA NUCLEAR SECURITY SERIES No. 44-T

DETECTION AT STATE BORDERS OF NUCLEAR AND OTHER RADIOACTIVE MATERIAL OUT OF REGULATORY CONTROL

TECHNICAL GUIDANCE

JOINTLY SPONSORED BY THE
INTERNATIONAL ATOMIC ENERGY AGENCY,
INTERNATIONAL CRIMINAL POLICE ORGANIZATION-INTERPOL,
UNITED NATIONS INTERREGIONAL CRIME AND JUSTICE
RESEARCH INSTITUTE,
UNITED NATIONS OFFICE OF COUNTER-TERRORISM,
UNITED NATIONS OFFICE ON DRUGS AND CRIME,
WORLD CUSTOMS ORGANIZATION

INTERNATIONAL ATOMIC ENERGY AGENCY
VIENNA, 2023

COPYRIGHT NOTICE

© IAEA, 2023

Printed by the IAEA in Austria
October 2023
STI/PUB/1952

IAEA Library Cataloguing in Publication Data

Names: International Atomic Energy Agency.
Title: Detection at state borders of nuclear and other radioactive material out of regulatory control / International Atomic Energy Agency.
Description: Vienna : International Atomic Energy Agency, 2023. | Series: IAEA nuclear security series, ISSN 1816–9317 ; no. 44-T | Includes bibliographical references.
Identifiers: IAEAL 22-01561 | ISBN 978–92–0–118621–8 (paperback : alk. paper) | ISBN 978–92–0–118721–5 (pdf) | ISBN 978–92–0–118821–2 (epub)
Subjects: LCSH: Radioactive substances. | Radioactive substances — Detection. | Radioactive substances — Security measures. | Radioactive substances — Safety measures. | Nuclear nonproliferation.
Classification: UDC 341.67 | STI/PUB/1952

FOREWORD

by Rafael Mariano Grossi
Director General

The IAEA Nuclear Security Series provides international consensus guidance on all aspects of nuclear security to support States as they work to fulfil their responsibility for nuclear security. The IAEA establishes and maintains this guidance as part of its central role in providing nuclear security related international support and coordination.

The IAEA Nuclear Security Series was launched in 2006 and is continuously updated by the IAEA in cooperation with experts from Member States. As Director General, I am committed to ensuring that the IAEA maintains and improves upon this integrated, comprehensive and consistent set of up to date, user friendly and fit for purpose security guidance publications of high quality. The proper application of this guidance in the use of nuclear science and technology should offer a high level of nuclear security and provide the confidence necessary to allow for the ongoing use of nuclear technology for the benefit of all.

Nuclear security is a national responsibility. The IAEA Nuclear Security Series complements international legal instruments on nuclear security and serves as a global reference to help parties meet their obligations. While the security guidance is not legally binding on Member States, it is widely applied. It has become an indispensable reference point and a common denominator for the vast majority of Member States that have adopted this guidance for use in national regulations to enhance nuclear security in nuclear power generation, research reactors and fuel cycle facilities as well as in nuclear applications in medicine, industry, agriculture and research.

The guidance provided in the IAEA Nuclear Security Series is based on the practical experience of its Member States and produced through international consensus. The involvement of the members of the Nuclear Security Guidance Committee and others is particularly important, and I am grateful to all those who contribute their knowledge and expertise to this endeavour.

The IAEA also uses the guidance in the IAEA Nuclear Security Series when it assists Member States through its review missions and advisory services. This helps Member States in the application of this guidance and enables valuable experience and insight to be shared. Feedback from these missions and services, and lessons identified from events and experience in the use and application of security guidance, are taken into account during their periodic revision.

I believe the guidance provided in the IAEA Nuclear Security Series and its application make an invaluable contribution to ensuring a high level of nuclear security in the use of nuclear technology. I encourage all Member States to promote and apply this guidance, and to work with the IAEA to uphold its quality now and in the future.

PREFACE

The IAEA Nuclear Security Series provides recommendations and guidance that States can use in establishing, implementing and maintaining their national nuclear security regimes.

IAEA Nuclear Security Series No. 15, Nuclear Security Recommendations on Nuclear and Other Radioactive Material out of Regulatory Control, provides recommendations to a State for the nuclear security of nuclear or other radioactive material that has been reported as being out of regulatory control, as well as for material that is lost, missing or stolen but has not been reported as such, or has been otherwise discovered. IAEA Nuclear Security Series No. 15 is jointly sponsored by the European Police Office (EUROPOL), the IAEA, the International Civil Aviation Organization (ICAO), the International Criminal Police Organization-INTERPOL (ICPO-INTERPOL), the United Nations Interregional Crime and Justice Research Institute (UNICRI), the United Nations Office on Drugs and Crime (UNODC) and the World Customs Organization (WCO).

The present publication provides more detailed guidance on meeting the recommendations set out in IAEA Nuclear Security Series No. 15. It addresses nuclear security detection systems and measures at State borders, with special consideration of designated points of exit or entry and border areas.

The present publication is jointly sponsored by the IAEA, the International Criminal Police Organization-INTERPOL (ICPO-INTERPOL), the United Nations Interregional Crime and Justice Research Institute (UNICRI), the United Nations Office of Counter-Terrorism (UNOCT), the United Nations Office on Drugs and Crime (UNODC) and the World Customs Organization (WCO).

CONTENTS

1. INTRODUCTION

BACKGROUND

1.1. The threat posed by nuclear and other radioactive material out of regulatory control is an important challenge that States face. The timely detection of this material can reduce the risk of its use in criminal or intentional unauthorized acts.

1.2. IAEA Nuclear Security Series No. 20, Objective and Essential Elements of a State's Nuclear Security Regime [1], identifies the establishment of systems and measures to detect nuclear security events as an essential element of a State's nuclear security regime.

1.3. Paragraph 5.6 of IAEA Nuclear Security Series No. 15, Nuclear Security Recommendations on Nuclear and Other Radioactive Material out of Regulatory Control [2], states:

> "Using the national threat assessment, the *competent authorities* should establish *nuclear security systems* for *detection* by instruments of nuclear and other *radioactive material* that is out of regulatory control. The *detection systems* should be based on a multilayered *defence in depth* approach and on the premise that such material could originate from both within or outside the State, and provide the necessary *detection* capability and capacity."

1.4. Paragraph 5.11 of Ref. [2] recommends the following:

> "[T]he State should continuously gather, store and analyse operational information with the goal of identifying any threat, suspicious activity or abnormality involving nuclear or other *radioactive material* that may indicate the intention to commit a criminal act, or an unauthorized act, with nuclear security implications".

1.5. IAEA Nuclear Security Series No. 21, Nuclear Security Systems and Measures for the Detection of Nuclear and Other Radioactive Material out of Regulatory Control [3], provides guidance for the development of an effective nuclear security detection architecture derived from a comprehensive, integrated detection strategy prepared by the State. Reference [3] states that "Effective border controls are critical in preventing and/or detecting the unauthorized transport of nuclear and other radioactive material."

1.6. This publication provides detailed guidance to supplement the guidance in Ref. [3]. States can use this guidance to design, implement and sustain effective nuclear security detection systems and measures at State borders that meet national nuclear security objectives and facilitate the efficient and effective movement of people, goods and conveyances. This publication also supplements the following Implementing Guides:

— IAEA Nuclear Security Series No. 37-G, Developing a National Framework for Managing the Response to Nuclear Security Events [4];
— IAEA Nuclear Security Series No. 24-G, Risk Informed Approach for Nuclear Security Measures for Nuclear and Other Radioactive Material out of Regulatory Control [5];
— IAEA Nuclear Security Series No. 31-G, Building Capacity for Nuclear Security [6].

1.7. This publication also complements the following Technical Guidance: IAEA Nuclear Security Series No. 34-T, Planning and Organizing Nuclear Security Systems and Measures for Nuclear and Other Radioactive Material out of Regulatory Control [7].

OBJECTIVE

1.8. The objective of this publication is to provide guidance to States on planning, implementing and evaluating systems and measures to detect at State borders nuclear and other radioactive material out of regulatory control.

1.9. This publication is intended to be used by national competent authorities and other organizations responsible for developing, designing, implementing and sustaining detection systems and measures at State borders, such as border protection authorities, customs authorities, national or local law enforcement agencies, regulatory bodies, national postal administrations and civil aviation authorities.

SCOPE

1.10. This publication addresses nuclear security detection systems and measures at State borders, with special consideration of designated points of exit or entry (POEs) and border areas. This publication does not address nuclear security detection systems and measures within a State.

1.11. The guidance in this publication applies to the detection at State borders of nuclear and other radioactive material out of regulatory control for all types of traffic flow involving people, goods and/or conveyances, including the following:

(a) All types of persons, including pedestrians, passengers, ship or airline crews, airport or seaport employees, or residents of border areas;
(b) All means of transport for people and cargo, including cars, vans, trains, buses, trucks, ships, boats, construction vehicles and conveyors;
(c) All types of goods, including personal items, luggage, mail, containerized shipments and bulk shipments.

1.12. This publication does not address response to a nuclear security event, guidance on which is provided in Ref. [4]. This publication therefore does not address response to nuclear or radiological emergencies involving nuclear or other radioactive material out of regulatory control at State borders, guidance on which is provided in Refs [8–14].

1.13. While reference is made in this publication to the need for radiation safety measures at the point of detection, such measures are not addressed in detail. Subsequent handling of seized materials is also outside the scope of this publication.

STRUCTURE

1.14. Section 2 provides information on the national detection strategy. Section 3 provides guidance on planning, implementing and evaluating detection systems and measures at State borders. Sections 4 and 5 focus on considerations specific to detection at designated POEs and in border areas, respectively. The Appendix provides descriptions of common types of detection equipment used at State borders. Annexes I and II present examples of content for the concept of operations, the design and standard operating procedures, and Annex III provides a detailed example of the process for evaluating alarms on declared shipments.

2. NATIONAL DETECTION STRATEGY AND NUCLEAR SECURITY DETECTION ARCHITECTURE

2.1. For the purpose of this publication, the term 'at a State border' is taken to mean one of the following:

(a)   At a designated POE within the State.
(b)   At an undesignated POE in a 'border area', that is, on the geographical line that separates a State from a neighbouring State, or in the area of the State lying along and close to this line. In the case of a border that crosses a lake or sea, the border area includes the area of water within the State between the border and the shore or coast, as well as the area of the State lying along and close to this shore or coast.

2.2. A designated POE is an officially designated place on the land border between two States, seaport, international airport or other point where travellers, means of transport, and/or goods are inspected. Often, customs and immigration facilities are provided at these POEs. Border areas include all locations at or near State borders and undesignated POEs, which include any air, land or water crossing point between two States that is not officially designated for travellers and/or goods by the State, such as green borders, seashores and local airports. The national detection strategy described in this publication guides the State's detection activities at designated POEs and in border areas.

LEGISLATIVE AND REGULATORY FRAMEWORK

2.3. Paragraph 3.2 of Ref. [2] states:

"As part of an overall framework, the State should establish and maintain effective executive, judicial, legislative and regulatory frameworks to govern the *detection* of and *response* to a criminal act, or an unauthorized act, with nuclear security implications involving any nuclear or other *radioactive material* that is out of *regulatory control*. Responsibilities should be clearly defined for implementing various elements of nuclear security and assigned to the relevant *competent authorities*".

2.4. Reference [2] also states (footnote omitted):

"3.3. In establishing legislative and regulatory frameworks to govern nuclear security, the State should define the conduct which it considers to be a criminal act, or an unauthorized act, with nuclear security implications.

"3.4. The State should establish criminal offences under domestic law which should include the wilful, unauthorized acquisition, possession, use, transfer or transport of nuclear or other *radioactive material* consistent with international treaties, conventions and legally binding United Nations Security Council resolutions.

"3.5. The State should also establish as criminal offences a threat or attempt to commit an offence as described in paragraph 3.4.

"3.6. The State should consider establishing as criminal offences, unlawful scams or hoaxes with nuclear security implications."

2.5. Moreover, para. 4.60 of IAEA Nuclear Security Series No. 29-G, Developing Regulations and Associated Administrative Measures for Nuclear Security [15], states:

"The State should include in its legislative and regulatory framework requirements for:

— A national strategy for the detection of criminal or intentional unauthorized acts with nuclear security implications involving nuclear or other radioactive material out of regulatory control;
— Nuclear security systems and measures for the detection of nuclear and other radioactive material out of regulatory control;
— Agreements for international cooperation and assistance in relation to the detection of nuclear and other radioactive material out of regulatory control."

2.6. Paragraph 4.61 of Ref. [15] states:

"The competent authorities with responsibilities for the detection of material out of regulatory control should be designated in primary legislation, such as nuclear law, national security legislation and legislation related to border protection and customs. The main competent authorities involved in the detection of material out of regulatory control include those with responsibilities to monitor and control the movements of goods and people. Competent authorities with responsibilities for the detection of material out of regulatory control may include the regulatory body, police and law enforcement agencies, customs authorities, border protection authorities and intelligence agencies."

2.7. Furthermore, para. 2.16 of Ref. [3] states:

"The legal framework should also provide the basis for the implementation of national import and export controls as well as customs and border operations for detection at designated and undesignated points of entry and/or exit (POEs), and at other strategic locations."

2.8. To facilitate inspections and detection, the competent authorities should have the appropriate authorization to stop, search and detain people and to seize goods and conveyances as part of their operations at the location where the inspections or detection take place.

2.9. Records of the calibration and maintenance of the equipment used in detection, and of the training of the front line officers[1], should be maintained. All activities should be conducted in accordance with established standards and certification specifications and should be documented, as this information might be needed as supporting evidence in a prosecution.

NATIONAL DETECTION STRATEGY

2.10. Paragraph 3.14 of Ref. [2] states:

"The State, through its coordinating body or mechanism, should inter alia:

— Ensure the development of a comprehensive national *detection* strategy based on a multilayered *defence in depth* approach within available resources".

---

[1] Front line officers are the responsible individuals from a designated government organization or institution who are first potentially in contact with nuclear and other radioactive material out of regulatory control, either through information alerts or instrument alarms.

2.11. Paragraph 2.8 of Ref. [3] states:

"The national detection strategy should determine the scope of, and priority assigned to the nuclear security detection architecture. It should articulate objectives for the detection systems and measures, and provide the basis for assignment of functions, including cooperation and coordination between the competent authorities and allocation of resources."

2.12. The possible movement of nuclear and other radioactive material along transit routes into and out of the State should be considered in the national detection strategy. When performing a national threat assessment, as described in para. 3.19 of Ref. [2], competent authorities should work closely together and consider "The threat through and to the transboundary movement and transport of goods and movement of persons".

2.13. States should consider applying a graded approach by prioritizing their designated POEs and border areas — taking into account factors such as the level of risk, the location and size of the POE or area, the volume and type of traffic through it, the cost of detection systems and measures and the strategic importance of the area — in order to implement more effective detection systems and measures in higher priority locations.

POLICY AND STRATEGY ATTRIBUTES OF A NUCLEAR SECURITY DETECTION ARCHITECTURE FOR STATE BORDERS

2.14. As described in Ref. [7], the first step of planning a nuclear security detection architecture involves the following:

"2.8. [P]lanners review the goals[2] of the detection architecture … and develop specific, measurable and actionable descriptions of activities that need to be completed to achieve these goals.

"2.9. These descriptions, referred to as functional outcomes[3], can be developed at different levels of specificity, and articulate specific directions for the design of the detection architecture …

[2] In [Ref. [7]] 'goals' refer to high level statements that set the general direction.
[3] In [Ref. [7]] 'functional outcomes' refer to specific descriptions of actions to be performed."

2.15. The goals and functional outcomes for the nuclear security detection architecture at designated POEs and in border areas should be based on the policy and strategy attributes described in para. 3.4 of Ref. [7]. The sustainability of the detection systems and measures, as well as the need for complementary measures to address the insider threat through mechanisms such as trustworthiness programmes, should also be taken into account when planning and implementing detection systems and measures at State borders.

3. PLANNING, IMPLEMENTING AND EVALUATING DETECTION SYSTEMS AND MEASURES AT STATE BORDERS

3.1. Developing and sustaining systems and measures for the detection at State borders of nuclear and other radioactive material out of regulatory control involves three phases: planning, implementation and evaluation. The State should identify a lead organization or coordinating body that will be responsible for each of these phases.

PLANNING DETECTION SYSTEMS AND MEASURES AT STATE BORDERS

3.2. During the planning phase, the competent authority responsible for detection at State borders should draft and finalize the concept of operations and the design for the detection systems and measures.

3.3. The concept of operations should define and describe the process by which information and equipment are to be used to detect nuclear or other radioactive material out of regulatory control and how the initial assessment is to be performed to determine whether an instrument alarm[2] or information alert[3] indicates the occurrence of an actual nuclear security event. An understanding of this process is needed to implement an effective design for nuclear security detection systems and measures that includes both equipment and operational aspects.[4]

3.4. The design is a document that describes in detail the resources that are needed, and where and when they are needed, to operate the detection systems and measures and hence implement the concept of operations. The design complements the concept of operations by identifying the physical location of detection instruments at designated POEs and in border areas, and by describing inspection locations, traffic flow and control mechanisms, and the locations of border personnel and supporting communications infrastructure (e.g. cameras, central and local alarm stations, command centres).

3.5. The concept of operations and the design should document the goals for the detection systems and measures at a specific site or area, should identify the relevant competent authorities and should designate the organizational roles and responsibilities, taking into account such information as the location of the detection systems and measures, operational scenarios and constraints. The concept of operations and the design should also describe the characteristics of a detection system in a manner that integrates operations, staffing, infrastructure and maintenance plans. The concept of operations and the design should be agreed on by relevant competent authorities and other organizations that are affected by the detection systems and measures.

3.6. The concept of operations and the design are important inputs to defining the specifications for equipment procurement and developing and finalizing the standard operating procedures. The concept of operations addresses the overall process for detection (i.e. who does what, when and where), including a high

---

[2] An instrument alarm is a signal from instruments that could indicate a nuclear security event requiring assessment. An instrument alarm may come from devices that are portable or deployed at fixed locations and operated to augment normal commerce protocols and/or in a law enforcement operation [2].

[3] An information alert is time sensitive reporting that could indicate a nuclear security event requiring assessment and may come from a variety of sources, including operational information, medical surveillance, accounting and consigner/consignee discrepancies, border monitoring, etc. [2].

[4] A concept of operations can be documented at the national level, describing multiple organizations' activities; at the organizational level, describing the activities within a single organization nationwide; or at the site level, describing activities for a specific site or area. This publication focuses on a site- or area-specific concept of operations.

level overview of the decision making that follows an alarm or alert. The design shows the physical deployment of equipment, identifies the information needed to support the concept of operations and provides instructions for traffic control, coordination of operations and information sharing (i.e. how personnel should carry out their assigned tasks). More details on the types of information included in standard operating procedures and how to develop such procedures for a designated POE or border area are provided in paras 3.69–3.75 and in Annex II.

3.7. When determining the level of detail that each document should include, States should consider each document's purpose and intended audience. For planning, for example, the concept of operations and the design might describe the general approach to how information will be shared, and with whom, and where equipment will be located. This information can help determine the number of staff needed and the type of equipment to be deployed. The standard operating procedures provide detailed instructions to the staff operating the equipment after its deployment on how information is to be shared among staff, which equipment should be used and where it should be located. Relevant equipment manuals might provide more detailed information than the standard operating procedures on the use and maintenance of the deployed equipment.

3.8. For example, the concept of operations might state that the safety of personnel should be considered when implementing detection systems and measures, while the standard operating procedures might dictate the use of personal radiation detectors by front line officers during all detection activities, and the equipment manual for the personal radiation detector normally provides specific instructions regarding, for example, how to turn on and read the detector. As another example, the concept of operations might state that all inbound trucks should be monitored by a radiation portal monitor, the standard operating procedures might describe which front line officer operates the radiation portal monitor workstation to process the alarms, and the workstation manual might provide detailed steps on using the software.

3.9. Examples of content for the concept of operations, for the design and for standard operating procedures are provided in Annexes I and II. States might find it useful to share experiences with other countries in order to understand good practices and common challenges related to their concepts of operations, design and standard operating procedures.

3.10. The development of the concept of operations and the design should be an iterative process, as they are complementary to each other. Detailed guidance on designating roles and responsibilities, drafting and finalizing the concept of operations and the design, and ensuring the sustainability of the system is provided in paras 3.11–3.59.

Designating roles and responsibilities

3.11. All competent authorities and other organizations with nuclear security responsibilities related to State borders should be identified in the concept of operations and the design. These organizations include those responsible for deploying or installing, operating or maintaining radiation detection equipment; conducting border protection and customs control; overseeing traffic control or the processing of people and cargo; operating a designated POE (e.g. an airport or port authority); providing technical expert support in the case of an instrument alarm or information alert; responding to a nuclear security event; transporting and storing detected material; and providing operational information such as intelligence.

3.12. The roles and areas of responsibility assigned to each organization should be documented. One or more organizations may be designated as front line organizations responsible for operating the detection systems and measures. Different organizations might carry out different types of inspection (i.e. of people, goods or conveyances) in different locations at the State borders (i.e. designated POEs or border areas), according to their specific areas of responsibility. National and local jurisdictions should be taken into account when identifying a front line organization and its roles and responsibilities. Some organizations with such responsibilities, such as luggage handling at airports or cargo handling at seaports, are not governmental agencies or regulatory bodies, but effective operations will rely on their cooperation and coordination with other agencies.

3.13. When designating roles and responsibilities for competent authorities and other organizations, the State should consider and address potential complicating factors, such as when the owner of the equipment is not the same organization that operates or maintains it, or when the authority responsible for assessing an alarm does not have jurisdiction in the area in which the detection equipment is located.

3.14. In particular, organizational responsibilities and jurisdictions should be considered when determining equipment locations. For example, if the customs organization is responsible for responding to radiation detection alarms and the detection equipment is located in a part of an international airport in which the customs organization has no jurisdiction, the concept of operations can be complicated by the need for additional responding organizations. In this situation, either the organizations should agree on a coordinated concept of operations or the detection equipment should be moved to a different location. A coordination agreement or memorandum of understanding can be used to clarify roles and responsibilities in cases where the concept of operations involves multiple organizations.

3.15. The organizations responsible for approving the design at different levels (e.g. site, local, national) should be clearly identified. An approval process for the design should be established during the planning phase, because the design can affect licensing and construction activities and equipment deployment during the implementation phase.

Drafting the concept of operations

3.16. The concept of operations (see paras 3.3–3.10) describes the activities expected to occur during the operation of detection systems and measures, including the initial assessment of alarms and alerts by front line officers. It might be informed by the specifications set out in the national detection strategy and the nuclear security detection architecture, as described in Ref. [3].

3.17. The concept of operations should describe the following:

(a) The goals and functional outcomes of the detection process;
(b) Any existing regulations, policies and constraints that might affect operations;
(c) Activities and decision making processes for detection, including the initial assessment and interactions between organizations;
(d) Assignment of responsibilities to competent authorities or other organizations for each activity in the detection process;
(e) Assignment of responsibilities for maintaining equipment and ensuring sustainability of the detection operations.

3.18. The concept of operations should also address how nuclear security detection measures are to be integrated into routine operations so as not to negatively affect other roles performed by competent authorities at borders. Competent authorities and other organizations with responsibilities for securing State borders manage a variety of complex situations within the context of many different national security objectives. A meeting with relevant representatives of all identified competent authorities and other organizations with responsibilities at a designated POE or in a border area should be held early in the development of the concept of operations to clarify and agree on respective roles and responsibilities.

3.19. The routine transport of nuclear or other radioactive material across State borders should be taken into account during the planning and development of the concept of operations. Nuclear and other radioactive material is widely used in authorized facilities and activities, including in the nuclear fuel cycle, in medical and industrial applications, and in agriculture and scientific research, and can be transported across borders between the locations of such facilities and activities. Several common products that are regularly transported for trade, such as fertilizers, building materials and ceramics, also contain naturally occurring radioactive material (NORM) [16], which is excluded or exempt from regulatory control. Detection instruments are likely to detect radiation from such material, and the concept of operations needs to allow for the prompt investigation of these alarms to rapidly determine that they are innocent alarms[5]. Detection instruments that identify the radionuclides emitting the radiation are needed to identify the source (or sources) of radiation and to determine whether an alarm is innocent or not (see para. 4.11). Identification of the radionuclides can also provide valuable information to competent authorities in cases where response measures are activated.

3.20. The concept of operations should also describe how information alerts will be processed by the relevant competent authorities. An information alert might provide information about nuclear or other radioactive material out of regulatory control entering or leaving a State through either designated POEs or border areas [3]. However, the validity of the information should be checked and procedures should be implemented to make sure sufficient information is gathered when acting on an information alert.

3.21. Protocols for sharing information should be taken into account in the development of the concept of operations, including sharing internally within a competent authority, between national agencies, between States, or with international organizations through established communication channels. For example, information exchange between States can be used to identify particular containers, conveyances and cargo that, or people who, might be carrying nuclear or other radioactive material out of regulatory control and thus warrant a greater degree of inspection.

---

[5] An innocent alarm is an alarm found by subsequent assessment to have been caused by nuclear or other radioactive material under regulatory control or exempt or excluded from regulatory control [17].

3.22. The concept of operations should describe the process that the front line officer should follow to determine whether an instrument alarm or information alert indicates the occurrence of a nuclear security event. The concept of operations should also describe the sequence of activities and decision points that should be worked through to reach this determination.

3.23. A systematic approach should be used for the detection of nuclear or other radioactive material out of regulatory control and the declaration of a nuclear security event, and the approach should be described in the concept of operations. The approach could include the following main stages:

(a)   Stage 1: Primary detection by instrument alarm and/or information alert.
(b)   Stage 2: Confirmation of the primary detection.
(c)   Stage 3: Confirmation of a radiation hazard.
(d)   Stage 4: Collection and analysis of information, including radionuclide identification.
(e)   Stage 5: Declaration of a nuclear security event and notification of the competent authorities, if indicated by the initial assessment.

3.24. The radiation protection and safety of the front line officer, and of any other people affected, should be considered during all stages.

3.25. The concept of operations might include text and flow charts to illustrate who does what and when. Figure 1 shows a flow chart of a generic process for detection (including the initial assessment), which could be tailored to specific designated POEs or border areas. Such a flow chart should clearly show each step of the decision making process.

3.26. The five stages listed in para. 3.23 can be adapted to integrate effectively with existing security processes at designated POEs or border areas, depending on the State's particular considerations and constraints. The five stages are described in more detail in paras 3.27–3.35.

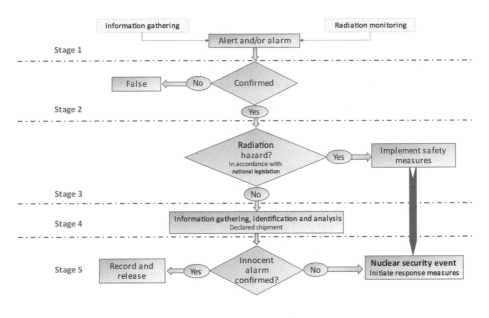

FIG. 1. Generic concept of operations flow chart.

Stage 1: Primary detection by instrument alarm and/or information alert

3.27. The process for detecting nuclear or other radioactive material out of regulatory control begins when an instrument alarm is triggered or a front line officer receives an information alert. Information alerts at a State border might originate from, for example, operational information obtained by front line officers, such as observations of suspicious behaviour by people crossing the border, the discovery of falsified or inaccurate documents, or data from complementary technologies (e.g. X ray scanning) showing inconsistencies between declared and actual items. Information alerts might also be received from other authorities within the State or from another State. An established procedure should be in place for receiving such information alerts and initiating a process for follow-up actions, depending on the type and credibility of the information (e.g. checks and controls at the border could be intensified). The subsequent stages of handling information alerts will be specific to each State's security arrangements, and therefore paras 3.28–3.35 on Stages 2–5 primarily address the handling of instrument alarms.

Stage 2: Confirmation of the primary detection

3.28. At Stage 2, the front line officer should confirm the validity of the primary detection, particularly to establish whether an alarm is false.[6] A false alarm could be caused, for example, by an equipment malfunction. The primary detection might be confirmed by more thoroughly screening with radiation detection equipment the person, item or conveyance that triggered the alarm.

3.29. After Stages 1 and 2, the front line officer should either release the entity that appeared to trigger the alarm (i.e. the person, item or conveyance) if the alarm is false, or investigate further (and proceed to Stage 3) if the alarm is confirmed.

Stage 3: Confirmation of a radiation hazard

3.30. At Stage 3, the front line officer should decide whether it is safe to proceed using the standard operating procedures developed according to national regulations.[7] If the front line officer determines it is unsafe to proceed to the next stage because of an actual or potential radiation hazard, the appropriate response organizations should be notified and appropriate protective actions and other response actions should be implemented (e.g. establishing an inner cordoned area and evacuating that area), according to the requirements and guidance in Refs [8–11].

Stage 4: Collection and analysis of information, including radionuclide identification

3.31. At Stage 4, if the alarm is confirmed and it has been determined that there is no radiation hazard and it is safe to proceed, the front line officer (or expert support personnel, depending on the agreed concept of operations) should gather and analyse the available information and conduct additional inspections, as necessary, to determine whether the alarm was an innocent alarm or an indication of a real nuclear security concern. Such inspections might vary considerably depending on the specific situation, and might include, for example, verification with a radionuclide identification instrument, examination of documentation, or other analysis and confirmation of information.

---

[6] A false alarm is an alarm found by subsequent assessment not to have been caused by the presence of nuclear or radioactive material [3].

[7] A gamma dose rate > 100 μSv/h at 1 m from the object or at 1 m above the ground is an indication of a possible radiological emergency (hazard), although dose limits may differ among Member States according to national regulations [13].

3.32. If the front line officer is presented with the documentation, signage, placards and labels required for a declared legal shipment of nuclear material or radioactive material, the front line officer or expert support personnel should consider the process outlined in Annex III for verifying declared shipments of nuclear or radioactive material with the use of specialized detection equipment.

Stage 5: Declaration of a nuclear security event and notification of the competent authorities, if indicated by the initial assessment

3.33. Using the information gathered and analysed in Stage 4, at Stage 5 the front line officer should proceed as follows:

(a)   If the alarm is determined to be innocent, the front line officer should release the entity that caused the alarm and record the event.
(b)   If the alarm is determined not to be innocent, the front line officer should notify the relevant competent authorities to initiate response procedures for a nuclear security event.
(c)   If the front line officer cannot determine the type of alarm, expert support should be obtained to assist with further information gathering and analysis.

3.34. If the dose rate measured or material found during an additional inspection presents an imminent danger to health and safety, or a security threat, a safe perimeter should be defined, the material should be secured and the relevant competent authorities should be notified. The concept of operations and the relevant response plans should document the roles and responsibilities and the process for these actions and notifications as appropriate.

3.35. Responsibilities for notification and for securing and placing material under regulatory control may be assigned to different organizations, or one organization may have multiple responsibilities. Actions to meet these responsibilities should be established in accordance with relevant response plans and procedures for nuclear security events.

Designing detection systems and measures at State borders

3.36. The design goals for detection systems and measures for designated POEs and border areas should be such as to achieve the objectives of the national detection strategy. The design should reflect the concept of operations, and the physical location of and timeline for each detection stage (as described in para. 3.23) should be clearly specified. For example, a design goal for an individual location might specify that the primary inspection will be conducted by front line officers using fixed detection instruments on all inbound traffic (i.e. people, goods and conveyances). As another example, a design goal might specify that, following an information alert, detection teams with mobile detection instruments will be deployed to patrol a specific area indicated by the alert. An initial survey of the areas to be covered might be necessary to develop the design.

3.37. The design should identify at least the following:

(a)   Traffic flow and control mechanisms;
(b)   Locations and nature of detection instruments;
(c)   Locations for inspections;
(d)   Checkpoints and patrol areas;
(e)   Data management and communications equipment, including how this equipment will communicate with the national communications system and whether it will be located in a central alarm station, local alarm station and/or server room;
(f)   Other infrastructure, such as traffic control equipment, to support the detection systems and measures.

3.38. The design should specify where people or conveyances that have been screened will be isolated while the initial assessment and any secondary inspection are conducted. Secondary inspections might include — in addition to gamma spectrometry — X ray scans or physical searches to check the contents of containers, conveyances or items of cargo, or to look for contraband in the possession of a person. A temporary holding or storage location should be identified in the design for cases in which further inspections or analysis are needed to assess an alarm or alert, taking into account the relevant safety and security requirements. Temporary holding or storage locations can also serve as material seizure points.

3.39. Paragraphs 3.40–3.48 provide more detail on the types of detection equipment and on the communications equipment and supporting infrastructure that could be included in the design.

Types of detection equipment

3.40. Radiation detection equipment can be used to detect, locate, measure and identify nuclear and other radioactive material. Some detection equipment can also record and store measured data, download data to a computer, or transmit data to a centre of operations or to technical expert support organizations. Some detection equipment is equipped with a global positioning system (GPS) and can be used for monitoring and mapping large areas. Radiation detectors can be portable, mobile or fixed.[8] Advanced spectroscopic detectors can identify specific radionuclides from the radiation they emit, although these detectors are rarely used for primary detection at borders owing to their cost. Detection equipment selected for deployment should be based on technically sound and proven technology.

3.41. The basic types of radiation detection equipment that can be used for detection purposes at designated POEs and in border areas are listed below and described further in the Appendix:

— Personal radiation detectors;
— Handheld gamma and/or neutron survey meters;
— Handheld radionuclide identification devices;
— Backpack based radiation detection systems (with or without radionuclide identification capability);
— Vehicle mounted radiation detection systems (with or without radionuclide identification capability);
— Fixed radiation portal monitors;
— Conveyor belt radiation monitors;
— Airborne radiation detection systems[9];
— Maritime radiation detection systems.

3.42. Each type of detection equipment has advantages and limitations: each has a particular sensitivity range and provides optimal detection under particular conditions. The design of the overall detection system should take into account the operating conditions for which the different types of equipment are particularly suited. For example, fixed equipment is particularly suitable for monitoring controlled traffic, as the detection sensitivity depends on conditions such as the speed of the conveyance being scanned and its distance from the instrument.

3.43. In practice, selecting equipment will involve balancing different considerations. For example, fixed installations typically provide the most sensitive detection and can scan the largest volume of traffic, but at a higher purchase, installation and maintenance cost than portable or mobile equipment. Fixed

---

[8] Equipment described as 'portable' can typically be carried and used by a person (e.g. a front line officer). Equipment described as 'mobile' can readily be moved from place to place, typically by means of a vehicle or other equipment, but is then fixed for use at the chosen location. Equipment described as 'fixed' cannot readily be moved.

[9] In this publication, airborne radiation detection systems are systems that move through the air, such as aircraft mounted systems.

equipment with radionuclide spectrum identification capabilities is even more expensive to purchase and maintain, but it can reduce the staffing needs relating to the conduct of secondary inspections. In addition, other infrastructure, such as protective barriers, might be needed to protect fixed detection equipment against damage, and traffic control measures might need to be installed. Mobile detection systems might have lower sensitivity but allow for greater operational flexibility in where, when and how they are deployed. Portable equipment might have even lower sensitivity, but it can be less expensive to deploy and maintain and can offer the highest degree of operational flexibility. In addition to cost and detection sensitivity, consideration should be given to related factors, such as staffing needs and sustainability.

3.44. Other factors to consider when determining which type of equipment to use and where to locate it include environmental conditions (e.g. temperature, wind speed and direction, dust levels, humidity, risks of flooding or lightning); estimates of traffic volumes and alarm rates; natural background radiation; the physical layout of the site; existing infrastructure (e.g. stable power supply, communications); possible interference from nearby inspection devices that emit radiation (e.g. non-intrusive gamma or X ray inspection devices); the physical space available for the secondary inspection and detention of people, goods or conveyances; and the feasibility of protecting equipment against damage, theft and sabotage.

Communications and supporting infrastructure

3.45. In addition to setting out the types and locations of detection equipment to be used, the design should specify how information will be communicated and presented to assist front line officers in making sound decisions during the detection process. For example, if information will need to be shared between detection equipment and local command centres, or between local offices and headquarters, the design should identify which communications system (e.g. computers, cameras, servers, software) and infrastructure (e.g. power supply, cabling) will be needed. The design should describe any regulatory or other requirements for storing, analysing and transmitting data generated by the radiation detection equipment, and how information will be collected, collated, retained and removed. The design might also specify more detailed procedures. For example, data from a fixed portal monitor, such as the measured count rate, should be sent over a network to the front line officer along with a camera image of the conveyance passing through the monitor, or closed channel communication devices should be available to patrol teams, with which they can send spectral data collected from a detection instrument over a wireless network to headquarters.

3.46. For detection systems in which multiple sets of equipment report to one central alarm station, the design should specifically describe the method for networking and should address factors such as the bandwidth needed, the number of locations to be connected and the distances between them.

3.47. The design should specify the infrastructure needed to support the detection instruments and their communications. For example, the installation of fixed equipment may involve foundations, protective bollards, a stable power supply with backup, and a means to identify objects causing instrument alarms, such as cameras with a video link. A detection system needs power and communications connections, whether or not it includes radiation portal monitors. Handheld and mobile detection equipment might need a constant power supply for operation or for regular recharging, and information from such devices might need to be downloaded to computers, for which a suitable connection mechanism might be needed.

3.48. The design should also specify the infrastructure for receiving information alerts and for protecting and sharing information. This should include computer security and information security measures, as required by the State or as necessary to prevent compromise of sensitive information or of computer based systems performing or supporting functions related to detection. For example, at some locations, information might be shared over dedicated communication channels, such as fibre optic cables or wireless networks. The relevant competent authority should evaluate the capabilities of potential adversaries and determine the extent to which detection and communications equipment needs to be protected and whether specific measures, such as the use of virtual private networks or encryption of communications, should be required.

Finalizing the concept of operations and the design

3.49. The concept of operations and the design should be formal documents[10] that are developed and approved by all appropriate competent authorities and other organizations with responsibilities related to detection. The participants in the development of the concept of operations and the design should include decision makers with the authority to commit their organizations to roles and responsibilities, including financial obligations, and operational staff who can ensure that the concept of operations and the design are operationally viable.

---

[10] They might be combined into one document that meets the objectives of both.

3.50. Before the concept of operations and the design are finalized, a site survey should be conducted at each planned operational location. A site survey is an activity by which a designated POE or border area is reviewed to confirm local conditions, validate plans and identify possible locations for radiation detection equipment, data processing and communications systems, and infrastructure.

3.51. The goal of the site survey is to collect information on and to document the following: the physical layout of the areas in which equipment is to be deployed; the standard vehicle, cargo and pedestrian traffic patterns for entry to and exit from the area; routes through the area for import, export, transit and transhipment; existing infrastructure, including communications systems and power supplies; information flow among competent authorities and other organizations with responsibilities related to detection (e.g. customs authority, border protection authority, port authority, site operator, site security); and local regulations and governing procedures.

3.52. A site survey for the installation of fixed detection equipment should be conducted by a survey team. The survey team should include the following individuals:

(a)   Technical experts capable of developing designs for civil engineering works, electrical systems and communications networks.
(b)   Engineers who can determine construction specifications and initiate construction planning.
(c)   Representatives of the organizations with responsibilities related to detection to facilitate discussion on issues such as how to respond to alarms at the site or in the patrolled area and how this process will influence decisions on the location of detection equipment.
(d)   Radiation detection experts who can perform a background radiation survey. The results of the radiation survey will inform the specifications for the design and the operation of the detection system.

3.53. The proposed concept of operations and the proposed design should ideally be tested with different scenarios to simulate the decision making process under different conditions (e.g. during a power failure, in extremely crowded conditions, during quiet shifts with few staff on duty, when secondary inspection areas are located far from the patrol area). This can help validate assumptions and identify gaps in the concept of operations and the design.

3.54. The formal meetings to revise and approve the concept of operations and the design could be facilitated by the competent authority with the lead responsibility for detection at the designated POE or in the border area. Decisions taken in such meetings should be approved by all parties and documented.

Sustainability planning

3.55. In accordance with para. 3.12(d) of Ref. [1], the competent authorities assigned responsibilities for nuclear security should be provided with "sufficient human, financial and technical resources to carry out the organization's nuclear security responsibilities on a continuing basis using a risk informed approach". National regulations, such as legal requirements for detection at State borders, can provide for continuing resource allocations for activities associated with the long term sustainability of detection systems and measures.

3.56. During the planning process, States should document their plans for the long term sustainability of the detection systems and measures, including resources for the equipment's life cycle and development of human resources, in order to ensure that the systems and measures remain effective over time. States should conduct periodic risk and threat assessments as a basis for determining whether the detection systems and measures remain appropriate and for identifying and implementing changes to the detection system when necessary.

3.57. Regular training schedules should be developed for the personnel who operate detection systems and measures, with attention given to the development of training materials and to the certification of the trainers.

3.58. The State should establish (a) a means by which detection equipment will continue to be maintained and repaired in the long term, including establishing responsibility for the equipment itself and for financing repair services, tools and spare parts, and (b) a plan for the training of maintenance and repair personnel. The relevant competent authorities should work with vendors, contractors and suppliers where necessary to understand and plan for equipment life cycle costs.

3.59. The sustainability plan should establish a method for assessing the performance of the detection system, including the performance of the equipment and of the staff. It should also include resources for a life cycle plan to address equipment ageing, obsolescence and eventual upgrade or replacement. Recurring costs, such as staffing and training costs, should be incorporated into the annual resources of the relevant competent authorities.

IMPLEMENTING DETECTION SYSTEMS AND MEASURES AT STATE BORDERS

3.60. The implementation phase includes the selection and procurement of equipment, its deployment and acceptance testing, and the training of personnel. The standard operating procedures should be drafted during this phase, indicating how the detection systems and measures will be operated at the designated POE and/or in the border area.

Equipment selection and procurement

3.61. Equipment specifications should be established as a basis for the selection and procurement of equipment. The choice of equipment should take account of the concept of operations and the design, national policy and guidance (e.g. the national threat assessment, national detection strategy, nuclear security detection architecture, established equipment performance requirements or regulations), and factors identified during sustainability planning.

Selection and procurement of radiation detection instruments

3.62. Radiation detection instruments should meet specifications designed to ensure performance consistent with established international and/or national standards.

3.63. Procurement decisions frequently involve finding a balance between functionality and cost. Procurement considerations relating to the functionality of the equipment (including associated computer hardware and software) include the following:

(a) Ability to support the concept of operations and the design;
(b) Ability to detect and measure radiation levels associated with materials of concern for nuclear security;
(c) Ability to identify such materials;
(d) Reliability (ability to consistently perform adequately) under expected environmental conditions at the detection location;
(e) Compatibility with existing equipment;
(f) Ability to meet specifications for the display, storage and retention of data;
(g) Ease and reliability of calibration;
(h) Certification as qualified equipment for the intended purpose;
(i) Training needs for operators;
(j) General ease of use.

3.64. Procurement considerations relating to cost include the following:

(a)   The initial purchase cost of the equipment itself;
(b)   The cost of deployment, including installation of communications systems and other infrastructure (e.g. systems for traffic control);
(c)   The cost of maintenance, repair and spare parts beyond those provided under warranty (which might depend on whether these parts are available locally);
(d)   The cost of ensuring the long term availability of spare parts;
(e)   The cost of installing and maintaining supporting infrastructure;
(f)   Staffing costs associated with operating and maintaining the equipment.

3.65. During selection and procurement, States might take into account any radiation detection equipment that already exists for other uses (e.g. safety, emergency preparedness and response) and that could also provide functions for nuclear security detection without compromising its primary purpose.

Communications systems

3.66. A radiation detection system typically includes a communications system to share alarm and alert information with operators, technical expert support, responders and others with relevant responsibilities. The elements of the communications system might also include methods to exchange information between competent authorities and to recover data after an unplanned shutdown or failure.

3.67. Verbal communication (e.g. by radio or telephone) is the most basic communications system. Alternatives include collecting and recording data digitally at a centralized location (e.g. at a central alarm station or national response centre) and sending data, or otherwise providing access to data, to whomever needs it (and is authorized to receive it). Hardware and software for communications, whether procured from vendors or suppliers or developed internally, should be carefully planned and resourced. Compatibility of data formats with existing and planned systems should be considered and, if necessary, used as a selection criterion. Some resource considerations include computer maintenance and the need for regular equipment updates. Security features of hardware and software should be specified in procurement contracts in accordance with the system design specifications.

Equipment deployment

3.68. Deployment of the selected equipment might involve civil engineering works as well as the receipt, acceptance, installation and configuration of the detection systems. Quality assurance measures should be applied to ensure that systems as deployed meet the performance specifications. Depending on the nature and scope of the deployment activities, specialized support might be needed to assist with developing and executing contracts, establishing power and communications connections, and ensuring the availability of office space and areas for the conduct of inspections.

Development of standard operating procedures

3.69. Standard operating procedures are detailed written documents that describe how competent authorities or other organizations should implement the systems and measures defined in the concept of operations and the design to ensure effective detection. For example, the standard operating procedures should set out in detail how front line officers determine whether an instrument alarm or information alert indicates a nuclear security event. Good standard operating procedures can help ensure that detection activities are conducted accurately, consistently, efficiently and effectively. An example of content for a standard operating procedure is provided in Annex II.

3.70. The standard operating procedures should be drafted after the concept of operations and the design have been finalized. The competent authority with the lead responsibility for detection at a designated POE or in a border area should have the primary responsibility for drafting the standard operating procedures in cooperation with any other organization that will use them. This competent authority should also be responsible for reviewing standard operating procedures after the implementation phase and updating them when needed (e.g. following changes to the site configuration). Standard operating procedures should be developed and maintained under formal version control procedures.

3.71. The number of standard operating procedures needed for the activities specified in the concept of operations will depend on such factors as the number of organizations involved in operating the detection systems, the types of standard operating procedure that these organizations already have in place, and whether relevant existing operating procedures can be modified to include detection related operating procedures. The standard operating procedures should reflect the complexity of the system (including the hardware and software involved), but they should provide clear and simple instructions for the intended users. It might be helpful to include schematic overviews (e.g. flow charts) that refer to other more detailed documents and to use existing formats that will be familiar to users. Standard operating procedures might address detection based on information alerts as well as instrument alarms, if appropriate, taking account of the type of designated POE or border area, the different operating steps and organizations involved in the detection systems and measures, and the different staff who might have roles in detection.

3.72. The standard operating procedures should be written in clear language tailored to the target audience; should avoid ambiguity; and should specify alternative courses of action only when absolutely necessary. They should maintain an appropriate balance between the need for detailed instructions and the need for the procedures to be of a manageable length. Standard operating procedures should also address the daily responsibilities of the operating staff at different levels of the hierarchical structure of the implementing authority.

3.73. Safety and radiation protection measures based on State regulations (general and specific to the types of activity) should be detailed in the standard operating procedures. Radiation protection measures should be consistent with relevant IAEA safety standards such as IAEA Safety Standards Series No. GSR Part 3, Radiation Protection and Safety of Radiation Sources: International Basic Safety Standards [18]. The standard operating procedures should include an up to date list of contacts (e.g. technical expert support, maintenance personnel, response team) to enable the front line officer to request support as needed, clearly indicating whom to contact and under what circumstances.

3.74. The standard operating procedures should list the locations of fixed detection equipment and describe the deployment of other equipment. They should contain clear instructions for the front line officer on how to recognize the type of alarm and what actions to take for each type, how and when to use instruments in the assessment of the alarm (including uploading and reporting data), and how and when to perform any necessary routine maintenance such as battery checks. They should (a) provide examples of alarm information displays and instructions for operating software and (b) describe actions necessary to control traffic and its speed and to move people, goods and conveyances to the secondary inspection area if they trigger an alarm. They might also include detailed techniques for searching people, goods and conveyances, as well as guidance on establishing isolation areas and on notifying competent authorities of a nuclear security event.

3.75. The standard operating procedures should specify the roles, responsibilities and actions of staff if an information alert is received and provide instructions for processing the alert and for notifying competent authorities and other organizations, as appropriate. The standard operating procedures should include an action plan or checklist to help in confirming whether the subject of an alert is likely to be carrying nuclear or other radioactive material out of regulatory control. The standard operating procedures should indicate which of the actions specified in the case of an instrument alarm are applicable in the case of an information alert.

Staffing for detection systems and measures

3.76. Sufficient staff, in number and expertise, are needed to operate, maintain and support detection systems and measures. Staffing decisions affecting front line officers and other key personnel at State borders should take into account the fact that locations with a large number of instrument alarms might need extra staffing to conduct secondary inspections and that it might be challenging to allocate many functions and priorities to a limited number of staff members.

3.77. Before and during the deployment period, competent authorities should identify staffing needs for operation, maintenance and technical expert support for detection systems and measures, and suitable staff should be employed or contractors engaged to meet these needs. Considerations in determining staffing needs should include, at a minimum, the number of personnel needed to perform specific tasks identified in the concept of operations and in the standard operating procedures; the qualifications, knowledge and experience necessary to fulfil the various roles and responsibilities; financial and other resource limitations; and the need to provide training and awareness programmes and conduct exercises to improve staff effectiveness and sustainability.

3.78. Other considerations in staff planning should include normal staff turnover and the need for continuity of knowledge, skills and abilities, as well as the need to adapt to changes in technology or procedures. Radiation detection should be part of the basic training curriculum for front line officers.

Human resource development

3.79. Human resource development during this phase should identify the staff members that need to be trained as well as their training needs to ensure the effective operation, maintenance and sustainability of detection systems and measures.

3.80. Types of training should be tailored to the roles and responsibilities of different staff members involved in detection and their different backgrounds. Front line officers should be trained in, at a minimum, the basics of radiation and radiation protection, awareness of nuclear security threats, the safe and effective use and maintenance of detection equipment, and relevant standard operating procedures. General awareness training for the staff of competent authorities and other relevant organizations could help to develop awareness of and support for detection activities. Training for maintenance staff or contractors will need to cover at least routine maintenance, repair and calibration of equipment. Technical expert support personnel should also be trained, as appropriate to their role, in supporting the detection systems and measures. In addition to training for new staff, periodic refresher training should be provided for existing staff.

3.81. After the detection systems and measures become operational, the relevant competent authority should consider organizing scenario based operational exercises, such as drills, tabletop exercises and field exercises, to support training and to assist in evaluating the detection systems and measures.

EVALUATING DETECTION SYSTEMS AND MEASURES AT STATE BORDERS

3.82. The performance of the detection systems and measures should be evaluated before they become operational and, thereafter, on a periodic basis. During the evaluation phase, the detection systems and measures should be tested to ensure their effectiveness and consistency with the concept of operations, the design and the standard operating procedures. The maintenance and training programmes should also be in place during this phase to support the operation of the detection systems and measures and human resource needs. A programme of exercises can be designed and conducted to evaluate the detection systems and measures.

System testing and evaluation

3.83. The State should specify the competent authority responsible for the testing and evaluation of the detection systems and measures. To ensure that they operate as planned, the detection systems and measures should be reviewed and subjected to operational and performance testing on a defined periodic basis. The concept of operations and the design should be used to develop the testing specifications and metrics that provide a documented basis for the evaluation of the detection systems and measures. The standard operating procedures should be used to evaluate the detection systems and measures and the operators to ensure that the operators have the necessary training and skills. It should also be assessed whether the standard operating procedures correctly reflect the detection systems and measures as deployed and whether they are adequate and are being followed.

3.84. Acceptance testing of detection equipment is normally conducted as part of the procurement process during the implementation phase. Equipment should not be formally accepted from the manufacturer or installer until it has passed acceptance tests to confirm that it meets the functional specifications for the detection system. During the evaluation phase, the same acceptance tests can be conducted again to confirm that the equipment remains functional. Regular equipment checks (i.e. operational testing) using check sources should be performed to confirm that the systems continue to respond as designed. Computer security measures for the communications and detection equipment, as well as protocols and procedures should also be evaluated.

3.85. Other types of evaluation, such as scenario based tabletop exercises or field exercises, can be conducted to test specific components or measures or the entire detection system. Tabletop exercises can help evaluate whether the concept of operations functions adequately under different circumstances when different organizations are involved. Field exercises can help evaluate the deployed detection systems and the planned operations. Exercises can be designed to focus on specific situations, depending on the evaluation goals, and may be scheduled or unannounced. For example, an unannounced exercise simulating a smuggling attempt can provide valuable insights into the normal workings at a designated POE. Careful planning to ensure the safety of all participants is particularly important for unannounced exercises. More detailed information on planning and conducting exercises can be found in Ref. [19].

Validation and revision of standard operating procedures

3.86. Changes to the detection systems and measures that affect the concept of operations and/or the design can be made during the implementation phase. Since the standard operating procedures should be based on the detection systems and measures as deployed, they should be modified if necessary and finalized after the systems and measures have been deployed, tested and exercised.

3.87. The concept of operations, the design and the standard operating procedures should be reviewed, and revised if necessary, if major modifications are made at the designated POE or in the border area, if changes are made to the detection systems and measures, if the threat environment changes, or if an exercise or real incident shows that the current arrangements are inadequate. Training programmes should be updated whenever the concept of operations, the design or the standard operating procedures have been modified.

4. CONSIDERATIONS FOR DETECTION SYSTEMS AND MEASURES AT DESIGNATED POINTS OF EXIT OR ENTRY

4.1. Conditions at designated POEs are different from those in border areas. Some differences in approach to planning, implementing and evaluating detection systems and measures are therefore needed. This section addresses some considerations that are specific to designated POEs.

4.2. Designated POEs are key points in the flow of people, goods and conveyances between States. In designing detection systems and measures for designated POEs, a likely priority is to avoid undue cost and inconvenience to legitimate business and travel. Careful planning can lead to systems and measures that are effective for nuclear security detection while minimizing negative impacts on the legitimate movement of people, goods and conveyances.

4.3. Measures should be in place to control and monitor the movement of people, goods and conveyances across the border at a designated POE, but the nature and extent of those controls will depend on the particular situation. Nuclear security detection systems and measures should be designed to integrate well with the existing border protection systems and should be consistent with those of State organizations involved in countering other types of trafficking.

INSPECTION OF LARGE VOLUMES OF PEOPLE, GOODS AND CONVEYANCES

4.4. Some designated POEs experience very large volumes of traffic. If a State's national detection strategy includes the radiation monitoring of all people, goods and conveyances crossing the border at designated POEs, without unacceptably impeding the legitimate movement of people, goods and conveyances, fixed radiation portal monitors or mobile systems with large volume detectors are often the only realistic options for large volumes of traffic. Suitable controls, such as barriers, traffic signals, speed bumps, railings or turnstiles, can be used to regulate the flow and speed of people and conveyances and can help to keep the general traffic moving, allowing only one person or conveyance at a time to pass by or through detection equipment and thereby satisfying specific speed and distance specifications for effective detection. Competent authorities should determine how alarms will be communicated to operating staff at the designated POE and whether people being monitored should be informed of the presence of the detection equipment.

4.5. The detection systems should allow front line officers to identify the person or conveyance that triggered an alarm and separate them from the rest of the traffic flow. For high traffic flows, technologies such as video cameras, optical character recognition or radiofrequency identification devices can help to meet this goal. The information from the detection instruments and the radionuclide identification devices can be integrated into a communications system that displays the combined information to front line officers, enabling them to locate the person or conveyance triggering the alarm. If this type of system is needed, the specifications — including the locations of front line officers' workstations — should be identified in the design.

4.6. If the system depends on the visual identification of a person or conveyance, the primary inspection relies on the constant involvement of the front line officers.

4.7. The system design should specify whether audio or visual alarms, or both, are to be used during the primary inspection. At some locations, the audio or visual indicators of the detection equipment can be turned off, and alarms only announced remotely to front line officers through a workstation, remote alarm panel or mobile phone.

4.8. Although primary detection at designated POEs is commonly in the form of instrument alarms from radiation portal monitors or the personal radiation detectors of front line officers, detection can also be triggered by information alerts based on observations of suspicious activity by front line officers. The concept of operations should specify how the validity of an alarm or alert is to be confirmed, taking account of the potential effects on day-to-day operations of the POE and on traffic flow and control measures. For example, if the front line officer seeks to confirm the validity of alarms at the location of the primary detection by redirecting people or conveyances back through a radiation portal monitor or by conducting additional secondary inspections using handheld equipment, other traffic will be impeded, which could result in long queues.

TARGETING CRITERIA FOR SCREENING

4.9. It might not be feasible to screen all traffic through a designated POE owing to resource constraints such as staffing and time available to respond to alarms or for other logistical reasons. In such cases, criteria should be developed and applied to select the people, goods and conveyances that will be monitored. These targeting criteria should be described in the concept of operations, and site specific procedures should be included in the standard operating procedures.

4.10. The targeting criteria should be risk informed and include consideration of factors such as the threat, the fraction of traffic that can reasonably be screened, and any supplementary information available about specific people, goods or conveyances. If targeting criteria are based on easily determined factors such as the destination or origin of cargo, an element of randomness should also be included to prevent adversaries from exploiting these criteria and avoiding detection.

LEGITIMATE SOURCES OF RADIATION

4.11. The concept of operations and the design should address the fact that the radiation detection instruments will produce innocent alarms (i.e. a real increase in radiation level is detected but is not due to inadvertent movement or trafficking of radioactive material). Innocent alarms typically occur at designated POEs as a result of shipments containing NORM, authorized shipments of radioactive material, and individuals who have recently undergone medical procedures involving radiopharmaceuticals. The expected overall rate of alarms should be estimated, and the necessary equipment, space and staffing should be provided to allow for the expected number of alarms to be adjudicated.

4.12. The detection systems and measures at designated POEs should allow the front line officers to confirm whether the source of detected radiation is legitimate and whether declared shipments of radioactive material are in compliance with established regulations and with the material declared (e.g. to confirm that no additional material is being trafficked within a legal shipment). Front line officers should be familiar with transport requirements for nuclear and other radioactive material [20].

4.13. Verifying whether a shipment declared to include radioactive material contains only the declared radionuclides and activity is a specialized task needing expertise and equipment that might not normally be present at a designated POE; it might involve isolation of the cargo while waiting for technical expert support.

States might therefore consider limiting the number of designated POEs through which such declared shipments can be imported or exported. A technique for confirming the contents of a declared shipment is described in Annex III.

ADDITIONAL DOCUMENTATION AVAILABLE FOR DECISION MAKING

4.14. A wide variety of supporting documentation is typically available to front line officers at designated POEs to assist with the assessment of alarms and the associated decision making process. Examples include shipping manifests, customs declarations and personal travel documents. The design should specify when and how the competent authority undertaking detection activities should have access to this information. Any detection event should be considered in the context of the relevant supporting documentation, taking account of the specific characteristics of the designated POE. For example, if a front line officer receives an instrument alarm from a cargo container, the shipping documents (including packing lists, bills of lading and invoices) should be available to the front line officer to consider whether NORM might be present, whether additional inspection is needed and whether the documentation contains any irregularities or inconsistencies.

ON-SITE STORAGE CONSIDERATIONS

4.15. Areas for additional inspections and for isolation and temporary storage of seized radioactive material should be identified in the planning phase and documented in the design. These areas should be selected taking into account the need to maintain safety and nuclear security without unduly impeding the flow of people and goods. Some designated POEs might be subject to regulations or other restrictions that prevent the storage of nuclear or radioactive material on the site. In such cases, the availability of a storage location might need to be addressed in the concept of operations or in the design.

NON-INTRUSIVE INSPECTION EQUIPMENT

4.16. If radioactive material or nuclear material is sufficiently shielded, it can pass through a radiation portal monitor without triggering an instrument alarm. The competent authority responsible for detection should consider screening cargo at designated POEs using non-intrusive inspection systems to detect shielding materials. Non-intrusive inspection equipment includes metal detectors and mobile or fixed X ray, gamma and backscatter imaging systems.

4.17. Non-intrusive inspection equipment can affect the performance of nearby radiation detectors such as radiation portal monitors, causing alarms or other interference. If such equipment is used at a designated POE that also uses radiation portal monitors, the non-intrusive inspection equipment and the monitors should be placed as far apart as possible. If close placement is unavoidable, a configuration should be chosen that minimizes the impact of background or interference radiation from the non-intrusive inspection equipment on the radiation portal monitor. Additional shielding (e.g. on an X ray machine) or the addition of a collimator to a radiation portal monitor can also reduce interference. Alternatively, procedures or software can be used to introduce alternate operating times of the X ray machines and the radiation detectors, to ensure that the detectors do not operate while X rays are being generated.

UPGRADES, CHANGES AND DAMAGE

4.18. Changes to the layout or operation of designated POEs might affect the detection systems. For example, the routes that people, goods and conveyances follow through the POE may be changed. If this can be foreseen during the planning phase, the design features in fixed detection equipment should be considered to facilitate the later relocation of the equipment, such as the use of overhead wiring rather than underground trenches, of above ground foundations that can be relocated, or of systems based on a mobile design. If not planned for, the relocation of fixed detection equipment can be expensive. It may therefore be preferable to delay the installation of fixed detection equipment until after the completion of planned site modifications or renovations.

4.19. The process for holding a person or company responsible for damaging equipment should be formalized and documented. The repair and replacement of detection equipment can be expensive, and mechanisms to assign liability to individuals who damage equipment should be implemented prior to operating the system.

4.20. Mobile detection systems can be used when a fixed detection system is temporarily unavailable or when the traffic exceeds the capacity of a fixed detection system. Such systems can also be used for secondary inspections, or as a means of primary detection when there is no fixed detection equipment in place. For example, mobile detection systems can be used to monitor movements of cargo at seaports.

5. CONSIDERATIONS FOR DETECTION SYSTEMS AND MEASURES IN BORDER AREAS

5.1. Nuclear security detection systems and measures in border areas should be integrated into the existing border security arrangements. Border areas normally include the State border and adjacent territory (excluding designated POEs). Compared with detection systems and measures for designated POEs, the detection systems and measures in border areas need to cover larger and more varied geographical areas without established checkpoints.

5.2. Border areas are often defined in national legislation to delineate areas where special border security measures are in place. One competent authority is typically assigned overall responsibility for border protection and law enforcement in the border area, but additional competent authorities may also have specific functions. Roles and responsibilities should therefore be clearly established and documented in the concept of operations.

5.3. The concept of operations and the design should describe planned ongoing operations in the border area, using an integrated approach to operations, intelligence and instrument deployment to detect nuclear and other radioactive material out of regulatory control. Because any violation in a border area is a cause for concern and could indicate the possibility of other violations, front line officers patrolling such an area should be prepared to apply established systems and measures for detecting nuclear or other radioactive material whenever they intercept anybody or anything crossing the border. To do this, they might have access to detection equipment at known locations in the field (e.g. at regional patrol posts) or they might carry such equipment with them, either routinely during regular patrols or only in response to an information alert.

5.4. Information gathering in border areas can be based on instrument measurements as well as the inspection of people, goods, conveyances and documents located in the border area. A clear procedure for confirming this information, including communication with technical expert support, if necessary, should be established and documented in the standard operating procedures. For example, if a person is found crossing the border without correct personal documentation, more information might be needed about the person, and the front line officer should be able to communicate with a shift supervisor or other personnel from the local border unit to obtain more information about the person.

LIMITED INFRASTRUCTURE AND HARSH ENVIRONMENTAL CONDITIONS

5.5. Any detection equipment deployed to a border area should be suitable for use in the environmental conditions of the border area. Some border areas have limited infrastructure to support detection systems and measures. For example, patrol posts might not have the regular power supply and climate control needed to support certain types of detection equipment. For border areas without an accessible and reliable power supply to recharge batteries, one of the available options is to use personal radiation detectors that operate with disposable batteries. Border areas might also have few roads and very limited communications infrastructure. In the case of an instrument alarm, for example from a personal radiation detector, further inspection may need to take place at a local or regional post where additional detection equipment is located, or mobile or portable equipment may need to be moved to where the primary detection occurred.

5.6. Detection equipment used in border areas should be sufficiently durable to remain reliable during normal patrol activities under conditions such as extreme weather and difficult terrain, and should be sufficiently tolerant of temperature fluctuations and other potentially challenging environmental factors. For example, detection systems for use in maritime environments should be waterproof and corrosion resistant. Equipment selection should also take into account the ability of border units to maintain the equipment in a field setting, considering such factors as long term robustness, battery life under local environmental conditions, and ease of maintenance and repair without specialized tools.

DETECTION OPERATIONS COVERING LARGE AREAS

5.7. In some border areas, very large areas of land or water need to be patrolled with limited detection resources. In such cases, the concept of operations and the design should describe how supporting equipment and expertise will be mobilized to support the investigation of an instrument alarm or information alert, taking account of the likely distance between the location of the alarm or alert and the equipment and expertise. Existing systems for general surveillance and visual observation of the border area may be used to deploy field personnel with detection instruments to the appropriate locations, and personnel with appropriate detection instruments and a means of transport may be based at existing premises in a border area, such as a local law enforcement office.

5.8. In large border areas without physical barriers to prevent border crossings, ground, airborne or waterborne patrols may exist on one or both sides of the border. Regional cooperation in the form of a memorandum of understanding or agreement between neighbouring States can provide an opportunity for developing shared patrols or complementary measures to increase the effectiveness of detection.

POPULATED BORDER AREAS

5.9. Border areas with significant numbers of permanent residents, nomadic populations, irregular migrants or refugees can present a particular challenge for detection. The border might be unclear or not well marked, and people in the area might cross the border repeatedly in the course of daily life, without being checked by border patrols. In such cases, detection measures should be integrated into existing border security procedures as needed.

COMMUNICATION CHALLENGES

5.10. In cases where the capabilities of primary detection equipment in border areas are limited (e.g. only personal radiation detectors carried by front line officers are available), more accurate instruments may be needed to determine whether the nuclear or other radioactive material detected by a front line officer exceeds regulatory thresholds. The standard operating procedures should include procedures for communicating with expert support (either stationed at another location or available to be deployed as a mobile expert support team) from the relevant competent authorities to assist front line officers in the decision making process. Detailed protocols should be established and followed when transmitting measurements and other data to the technical expert support organization.

GEOGRAPHY AND TERRAIN

5.11. Depending on the geography of the border area, there may be more than one competent authority responsible for border security. For example, border guards might be responsible for border security at land borders, while coastguards might have that responsibility at water borders or on the shore near such borders. In such cases, the competent authorities should cooperate to provide consistently effective detection capabilities throughout the border areas.

5.12. Land border areas can be difficult to patrol owing to hostile terrain or environment, the presence of vegetation, and/or the large distances to be covered. Available detection systems might be limited to personal radiation detectors or vehicle based systems.

5.13. Detection systems deployed at maritime borders may include vessel based systems for detecting nuclear or other radioactive material in boats, or personal radiation detectors or other handheld instruments to be used by coastguard personnel when boarding boats. Airborne radiation detection systems are rarely used at borders, except during targeted searches.

5.14. Radiation screening of aircraft passengers and cargo can be performed at airports: international flights arrive at designated POEs and domestic flights land in the State's interior, not in a border area. Aircrafts illegally crossing the border should be detected by local border authorities and handled as border violators, but radiation detection should be included as part of the follow-up investigation.

Appendix

EQUIPMENT FOR RADIATION DETECTION AT STATE BORDERS

A.1. Instrument alarms can originate from a wide variety of radiation detection instruments. This Appendix describes different types of equipment that are typically used for radiation detection at State borders. Some are small enough to be worn (e.g. personal radiation detectors), some are handheld or carried as a backpack, and some are vehicle based. They also differ by function: some are used to detect radiation from radioactive material, some are used to locate the material more precisely after detection of the radiation, and some are used to identify radionuclides. More information about how to select an equipment type and model can be found in Ref. [21].

A.2. Personal radiation detectors are pocket sized, lightweight radiation detectors that can be worn on the body for the rapid detection of gamma and sometimes neutron radiation. These instruments give an alarm (audio, visual or vibrating) if the measured radiation level exceeds a preset threshold and they are generally intended to provide notification of potentially unsafe conditions. They are used to ensure personal safety with little or no intrusion or disruption of activities. They are used primarily by front line officers (e.g. border guards, coastguards, customs officers, law enforcement teams) as they are small, compact and user friendly; can be operated in extreme environmental conditions; and minimal training is needed to operate them. The wearer should be able to use the detector effectively while performing other tasks. They are the least expensive type of radiation detection equipment but have limited sensitivity.

A.3. Handheld gamma and/or neutron survey meters are portable radiation detectors used to search for and locate nuclear and other radioactive material. They are larger than personal radiation detectors and generally offer higher sensitivity, though lower than radiation portal monitors.

A.4. Handheld radionuclide identification devices are radiation detectors that can also collect and analyse the energy spectrum emitted by radionuclides and provide radionuclide identification. They may also contain a neutron detector for indicating the presence of neutron radiation. They have built-in software for spectral analysis and contain libraries of radionuclide data, making them capable of identifying the radionuclides most commonly encountered by front line officers. The main desired characteristics of radionuclide identification devices are sensitivity to gamma radiation, reliability of radionuclide identification, and approximate exposure rate indication. When radiation sources are detected by screening devices such as radiation portal monitors or personal radiation detectors, radionuclide identification devices can be used for secondary inspections to determine the source of radioactivity and evaluate the potential threat. Most radionuclide identification devices can also be used as handheld gamma and/or neutron survey meters to locate the source of radiation.

A.5. Backpack based radiation detectors are instruments where the detector (gamma and/or neutron, with or without identification capabilities) and associated electronics are contained in a backpack to be carried by the user for executing discreet searches in public areas. They are particularly useful for radiation surveys of large areas before or during major public events, or for detecting radiation in close proximity (e.g. while walking down the centre of a passenger train or bus). They can also be used temporarily for area monitoring or can be mounted on a small vehicle. The systems may be equipped with a GPS for mapping purposes. Important considerations for their use are weight, ergonomics, battery life and charge time, training time, ease of use, and capability for data transmission.

A.6. Vehicle mounted radiation detection systems are mobile radiation detection systems that are mounted on or inside a vehicle and may also be referred to as 'mobile detection systems'. They may be able to measure gamma and/or neutron radiation and may incorporate identification of gamma emitting radionuclides. They may be equipped with a GPS and provide search and localization capabilities. Operationally, they may be used in motion or as stationary equipment, so they offer increased flexibility.

A.7. Fixed radiation portal monitors are pass-through, non-intrusive monitors consisting of one or two pillars containing gamma radiation detectors, in some cases complemented by neutron detectors when sensitivity to nuclear material is desired. They can be used to screen pedestrians, vehicles, packages, personal luggage and other cargo. If the radiation measurement exceeds a preset threshold, the radiation portal monitor will produce an alarm to indicate the presence of nuclear or radioactive material. Systems include an occupancy sensor and may be linked to a means of video recording. Fixed radiation portal monitors are often deployed to monitor traffic at checkpoints and at designated POEs such as seaports, airports, land border crossings, rail crossings and international mail facilities. They are highly sensitive but expensive to procure and install. Spectroscopic radiation portal monitors can both detect radiation and identify the radionuclides, but these are more expensive to procure, install and maintain than standard radiation portal monitors.

A.8. Conveyor belt radiation monitors are portal monitors for the operation of which the material is put through the detectors by means of a conveyor drive, making them suitable for monitoring large quantities of items. A specific application can be found in monitoring public mail, where parcels and letters are placed on a conveyor belt for high sensitivity detection of gamma and neutron radiation; the monitors may be combined with X ray screening systems.

A.9. Airborne radiation detection systems can be mounted inside or outside an aircraft, including unmanned aerial vehicles. They may be used for measurement, detection and localization of radioactive material, and the data obtained by these systems are typically used for area mapping. They may be able to measure gamma and/or neutron radiation and may incorporate identification of gamma emitting radionuclides.

A.10. Maritime radiation detection systems can be mounted on or placed inside a maritime vessel. They may be operated in motion or in stationary mode. They may be able to measure gamma and/or neutron radiation, may incorporate identification of gamma emitting radionuclides and may be equipped with a GPS. They are manufactured for operation in marine environments.

REFERENCES

[1] INTERNATIONAL ATOMIC ENERGY AGENCY, Objective and Essential Elements of a State's Nuclear Security Regime, IAEA Nuclear Security Series No. 20, IAEA, Vienna (2013).

[2] EUROPEAN POLICE OFFICE, INTERNATIONAL ATOMIC ENERGY AGENCY, INTERNATIONAL CIVIL AVIATION ORGANIZATION, INTERNATIONAL CRIMINAL POLICE ORGANIZATION-INTERPOL, UNITED NATIONS INTERREGIONAL CRIME AND JUSTICE RESEARCH INSTITUTE, UNITED NATIONS OFFICE ON DRUGS AND CRIME, WORLD CUSTOMS ORGANIZATION, Nuclear Security Recommendations on Nuclear and Other Radioactive Material out of Regulatory Control, IAEA Nuclear Security Series No. 15, IAEA, Vienna (2011).

[3] INTERNATIONAL ATOMIC ENERGY AGENCY, Nuclear Security Systems and Measures for the Detection of Nuclear and Other Radioactive Material out of Regulatory Control, IAEA Nuclear Security Series No. 21, IAEA, Vienna (2013).

[4] INTERNATIONAL ATOMIC ENERGY AGENCY, Developing a National Framework for Managing the Response to Nuclear Security Events, IAEA Nuclear Security Series No. 37-G, IAEA, Vienna (2019).

[5] INTERNATIONAL ATOMIC ENERGY AGENCY, INTERNATIONAL CRIMINAL POLICE ORGANIZATION-INTERPOL, Risk Informed Approach for Nuclear Security Measures for Nuclear and Other Radioactive Material out of Regulatory Control, IAEA Nuclear Security Series No. 24-G, IAEA, Vienna (2015).

[6] INTERNATIONAL ATOMIC ENERGY AGENCY, Building Capacity for Nuclear Security, IAEA Nuclear Security Series No. 31-G, IAEA, Vienna (2018).

[7] INTERNATIONAL ATOMIC ENERGY AGENCY, Planning and Organizing Nuclear Security Systems and Measures for Nuclear and Other Radioactive Material out of Regulatory Control, IAEA Nuclear Security Series No. 34-T, IAEA, Vienna (2019).

[8] FOOD AND AGRICULTURE ORGANIZATION OF THE UNITED NATIONS, INTERNATIONAL ATOMIC ENERGY AGENCY, INTERNATIONAL CIVIL AVIATION ORGANIZATION, INTERNATIONAL LABOUR ORGANIZATION, INTERNATIONAL MARITIME ORGANIZATION, INTERPOL, OECD NUCLEAR ENERGY AGENCY, PAN AMERICAN HEALTH ORGANIZATION, PREPARATORY COMMISSION FOR THE COMPREHENSIVE NUCLEAR-TEST-BAN TREATY ORGANIZATION, UNITED NATIONS ENVIRONMENT PROGRAMME, UNITED NATIONS OFFICE FOR THE COORDINATION OF HUMANITARIAN AFFAIRS, WORLD HEALTH ORGANIZATION, WORLD METEOROLOGICAL ORGANIZATION, Preparedness and Response for a Nuclear or Radiological Emergency, IAEA Safety Standards Series No. GSR Part 7, IAEA, Vienna (2015).

[9]   FOOD AND AGRICULTURE ORGANIZATION OF THE UNITED NATIONS, INTERNATIONAL ATOMIC ENERGY AGENCY, INTERNATIONAL LABOUR OFFICE, PAN AMERICAN HEALTH ORGANIZATION, WORLD HEALTH ORGANIZATION, Criteria for Use in Preparedness and Response for a Nuclear or Radiological Emergency, IAEA Safety Standards Series No. GSG-2, IAEA, Vienna (2011). (A revision of this publication is in preparation.)

[10]   FOOD AND AGRICULTURE ORGANIZATION OF THE UNITED NATIONS, INTERNATIONAL ATOMIC ENERGY AGENCY, INTERNATIONAL LABOUR OFFICE, PAN AMERICAN HEALTH ORGANIZATION, UNITED NATIONS OFFICE FOR THE COORDINATION OF HUMANITARIAN AFFAIRS, WORLD HEALTH ORGANIZATION, Arrangements for Preparedness for a Nuclear or Radiological Emergency, IAEA Safety Standards Series No. GS-G-2.1, IAEA, Vienna (2007). (A revision of this publication is in preparation.)

[11]   FOOD AND AGRICULTURE ORGANIZATION OF THE UNITED NATIONS, INTERNATIONAL ATOMIC ENERGY AGENCY, INTERNATIONAL CIVIL AVIATION ORGANIZATION, INTERNATIONAL LABOUR OFFICE, INTERNATIONAL MARITIME ORGANIZATION, INTERPOL, OECD NUCLEAR ENERGY AGENCY, UNITED NATIONS OFFICE FOR THE COORDINATION OF HUMANITARIAN AFFAIRS, WORLD HEALTH ORGANIZATION, WORLD METEOROLOGICAL ORGANIZATION, Arrangements for the Termination of a Nuclear or Radiological Emergency, IAEA Safety Standards Series No. GSG-11, IAEA, Vienna (2018).

[12]   INTERNATIONAL ATOMIC ENERGY AGENCY, Method for Developing Arrangements for Response to a Nuclear or Radiological Emergency, EPR-Method 2003, IAEA, Vienna (2003).

[13]   INTERNATIONAL ASSOCIATION OF FIRE AND RESCUE SERVICES, INTERNATIONAL ATOMIC ENERGY AGENCY, PAN AMERICAN HEALTH ORGANIZATION, WORLD HEALTH ORGANIZATION, Manual for First Responders to a Radiological Emergency, EPR-First Responders 2006, IAEA, Vienna (2006).

[14]   INTERNATIONAL ATOMIC ENERGY AGENCY, Generic Procedures for Assessment and Response during a Radiological Emergency, IAEA-TECDOC-1162, IAEA, Vienna (2000).

[15]   INTERNATIONAL ATOMIC ENERGY AGENCY, Developing Regulations and Associated Administrative Measures for Nuclear Security, IAEA Nuclear Security Series No. 29-G, IAEA, Vienna (2018).

[16]   EUROPEAN POLICE OFFICE, INTERNATIONAL ATOMIC ENERGY AGENCY, INTERNATIONAL POLICE ORGANIZATION, WORLD CUSTOMS ORGANIZATION, Combating Illicit Trafficking in Nuclear and Other Radioactive Material, IAEA Nuclear Security Series No. 6, IAEA, Vienna (2007).

[17]   INTERNATIONAL ATOMIC ENERGY AGENCY, INTERNATIONAL CRIMINAL POLICE ORGANIZATION-INTERPOL, UNITED NATIONS INTERREGIONAL CRIME AND JUSTICE RESEARCH INSTITUTE, Radiological Crime Scene Management, IAEA Nuclear Security Series No. 22-G, IAEA, Vienna (2014).

[18] EUROPEAN COMMISSION, FOOD AND AGRICULTURE ORGANIZATION OF THE UNITED NATIONS, INTERNATIONAL ATOMIC ENERGY AGENCY, INTERNATIONAL LABOUR ORGANIZATION-INTERPOL, OECD NUCLEAR ENERGY AGENCY, PAN AMERICAN HEALTH ORGANIZATION, UNITED NATIONS ENVIRONMENT PROGRAMME, WORLD HEALTH ORGANIZATION, Radiation Protection and Safety of Radiation Sources: International Basic Safety Standards, IAEA Safety Standards Series No. GSR Part 3, IAEA, Vienna (2014).

[19] INTERNATIONAL ATOMIC ENERGY AGENCY, Preparation, Conduct and Evaluation of Exercises for Detection of and Response to Acts Involving Nuclear and Other Radioactive Material out of Regulatory Control, IAEA Nuclear Security Series No. 41-T, IAEA, Vienna (2020).

[20] INTERNATIONAL ATOMIC ENERGY AGENCY, Regulations for the Safe Transport of Radioactive Material, 2018 Edition, IAEA Safety Standards Series No. SSR-6 (Rev. 1), IAEA, Vienna (2018).

[21] INTERNATIONAL ATOMIC ENERGY AGENCY, Technical and Functional Specifications for Border Monitoring Equipment, IAEA Nuclear Security Series No. 1, IAEA, Vienna (2006).

Annex I

EXAMPLES OF CONTENT FOR THE CONCEPT
OF OPERATIONS AND THE DESIGN

I–1. The purpose of the concept of operations and the design is to describe how the radiation detection system will accomplish the State's strategy for detection systems and measures at a State border. The concept of operations and the design may be one document or a set of documents, depending on the size and scope of the radiation detection system deployed. These documents can be used as a basis for deploying radiation detection equipment, developing detailed system design specifications for system installation (as applicable), and planning procedures for agencies involved in detection and initial assessment operations at designated points of exit or entry (POEs) or in border areas.

I–2. The descriptions that follow provide examples of the type of information that could be included in the concept of operations and the design.

CONTENT FOR THE CONCEPT OF OPERATIONS

Background information

I–3. This section includes a summary of the key elements of the nuclear security detection systems and measures and the relevant national detection strategy or policy, as follows:

(a) Summary of the identified threats, general vectors and pathways that are of concern to the State;
(b) The competent authorities and other stakeholders that will support the detection system at each site or border area, including any pertinent information from national nuclear security response plans;
(c) The legislative basis, defined as the existing legislation and regulations for nuclear and other radioactive material out of regulatory control, international conventions, and any other legislation in the State applicable to the detection of criminal or intentional unauthorized acts involving these materials (e.g. illicit trafficking, malicious acts).

Objectives and goals

I–4. This section includes the overall objectives and goals of the State's detection systems and measures, as well as relevant details on equipment performance specifications, roles and responsibilities of the competent authorities and other relevant organizations, and prioritization of sites and patrol areas.

Information specific to the designated point of exit or entry or border area

I–5. This section documents conditions at the deployment location that might impact operations or otherwise affect the deployed system.

Prioritization considerations

I–6. This section describes the features that make this designated POE or border area an important location for detection and its status in the State's overall prioritization.

Objective and goals of detection operations

I–7. This section describes what the detection operations need to accomplish and how it will be accomplished. This information is based on a threat or risk assessment that might or might not be defined in this document. An example is shown in Box I–1.

BOX I–1: NOTIONAL EXAMPLE OF OBJECTIVES AND GOALS FOR DETECTION OPERATIONS AT AN AIRPORT

Objective:

Detection capabilities will be enhanced to ensure that nuclear and other radioactive material out of regulatory control will be identified.

Goals:

- The airport will scan [X amount of] cargo, mail, priority mail and passenger luggage.
- The [agency] operating staff will be trained to locate, isolate and identify nuclear or other radioactive material out of regulatory control.
- The equipment will be maintained in accordance with the manufacturer's specifications.

Location

I–8. This section describes the location of the site, including latitude and longitude, and includes references to other known locations (e.g. distance and direction from a known city) and other relevant information.

Site or area characteristics and layout

I–9. This section includes a detailed layout of the site that shows major site features, site size, distances, site orientation and geography and provides information on existing operations at the site.

I–10. Annotated maps, drawings and satellite images are helpful visuals to include in this section.

Trade and commodities (for designated points of exit or entry)

I–11. This section describes, if applicable, any trade that occurs at the site. It includes information on the volume of trade, on the partners with which the trade occurs, and on the types of commodity that most frequently pass through the site. This information is important since many commodities contain naturally occurring radioactive material, which can affect the way a radiation detection system operates in certain environments and impact the expected alarm rate.

I–12. The section also includes details on the types and volumes of commodities that are generally imported and exported or transhipped (if applicable). These details might impact the operational design and staffing needs at the site.

Traffic types and throughput (for designated points of exit or entry)

I–13. This section focuses on the types of traffic (e.g. pedestrian, personal vehicles, trucks, luggage) that pass through the site and on the gates, paths, entries and exits that each type of traffic uses. This section also describes which types of traffic the detection system will cover and which types of traffic it will not cover. If jurisdiction differs according to traffic type, this needs to be explained.

I–14. The section also includes the overall throughput and direction (e.g. import, export, transhipment) for each type of traffic to assess the detection needs to cover the different types of traffic and to handle the expected volume for each type of traffic.

Background radiation assessment (for fixed equipment at designated points of exit or entry)

I–15. This section contains the results of the background radiation assessment that is needed before selecting and installing any fixed radiation detection system. This assessment is performed using a radiation survey meter to record the gross counts in each particular area in which radiation detection equipment will be used.

Site or border area operations

I–16. This section includes details of specific site or area conditions and existing and proposed operations.

Roles and responsibilities

I–17. This section identifies the organizations that will have a role in the deployed system and defines the responsibilities of each. This includes identifying the organization that is responsible for the site and for detection system operation and sustainability. The section also identifies the agency (or agencies) that will own, operate, maintain and repair the radiation detection equipment, including all associated infrastructure, hardware and software, and train others on its use.

I–18. Different organizations may be responsible for different areas. For border areas, this section details which organizations patrol which areas and under what conditions, as well as what their specific responsibilities are.

I–19. For designated POEs, this section may include organizations that have roles and responsibilities for the site's property and may provide the following information:

(a) Site owner: the organization that owns the site and the site's property.
(b) Site operator(s): the organizations responsible for day-to-day site operations, including official duties, maintenance and site security.
(c) Site change authorities: the organizations that have authority on changes to the site.

Technical expert support agency

I–20. If a front line officer needs technical assistance to assess an alarm of concern (i.e. radionuclide identification) or to recover material and place it under regulatory control, scientific expertise from a relevant competent authority needs to be made available. This section identifies the technical expert support organization and the mechanism for communicating with that organization.

Incident or emergency response organizations

I–21. If dangerous material is found or a nuclear or radiological emergency is declared at the detection location, response expertise from the relevant competent authorities needs to be made available. This section identifies the response organizations or refers to relevant national response plans or the national radiation emergency plan.

Decision process at a designated point of exit or entry or border area

I–22. This section is used to describe the specific details of the concept of operations to be used at each location or border area identified for the deployment of a radiation detection system. The concept of operations for each location needs to be completed and communicated to the operators before finalizing the design for the radiation detection system.

I–23. The general process to be followed by the front line officer is described in para. 3.23 of the main text.

I–24. Following the guidelines outlined in this annex, specific locations can be identified where detection equipment will be deployed, and it will be indicated whether that equipment will be placed in a fixed location or used in a certain geographical region. All potential pathways of radioactive material are analysed, and a concept of operations is developed to detect material along each pathway. The concept of operations will identify where exactly along the pathway radiation detection equipment and detection measures are to be deployed or used.

I–25. For designated POEs, such as seaports, airports and land border crossings, a concept of operations approach is defined for each potential pathway through the designated POE. Potential pathways may include the following:

(a) Entry or import (pedestrians, vehicles, containerized cargo);
(b) Exit or export (pedestrians, vehicles, containerized cargo);
(c) Transhipments (cargo, luggage);
(d) Rail (containerized cargo).

I–26. For other border areas, a concept of operations approach considers the patrol areas and possible pathways.

Operational scenarios

I–27. This section includes the range of circumstances that might be encountered and a description of how the concept of operations can be applied. This helps validate the concept of operations and provides instruction to the participating organizations on how they will be expected to operate and interact under different scenarios. The concept of operations and the operational scenarios will be used in identifying and developing the standard operating procedures presented in Annex II.

CONTENT FOR DETECTION SYSTEM AND MEASURES DESIGN

I–28. The detection system and measures design describes the layout, infrastructure and operations of each site where radiation detection equipment will be installed. The design supports and documents the decisions made concerning the type and location of equipment to be installed at the site, conveys the conceptual design for equipment locations or deployments, and defines the specifications for installation of the equipment. The design will typically include both written installation and equipment specifications and conceptual drawings of the site.

Design specifications (by area, pathway or lane)

I–29. This section defines the detailed specifications necessary to ensure a properly functioning and performing detection system. The system needs to remain functional, and it also needs to perform properly to meet the detection goals under existing operational constraints and other factors of the site.

Pillar spacing (for fixed radiation portal monitors)

I–30. This section defines the pillar spacing for fixed radiation portal monitors. Pillar spacing is based on equipment type, manufacturer recommendations, background radiation, operational considerations and performance specifications (e.g. sensitivity of the monitors, amount of material the monitors can detect).

Physical protection and traffic controls

I–31. This section defines which physical protection measures and traffic controls are needed to adequately protect personnel and the detection equipment and to maintain traffic flow. Accommodations need to be made for oversized traffic that will not fit through the radiation portal monitors. Examples of traffic controls include bollards, barriers, railings and turnstiles.

Environmental controls

I–32. This section defines what environmental controls are necessary to ensure equipment operation (e.g. air conditioning for the server room). The design needs to take into account seasonal variations in the weather and mitigation methods (e.g. shelters, wind barriers) for both the public being scanned by the equipment and the front line officer operating it.

Power and communications infrastructure

I–33. This section defines the specifications for equipment power and communications connectivity. It needs to be determined whether existing conduits and media (e.g. copper, fibre optic) can be used or whether new equipment needs to be installed. During the site survey, the locations of power and communications conduits and access points need to be identified, as well as the distances between connection points if new cabling or underground trenching is needed.

Communications system

I–34. Fixed or vehicle based radiation detection systems often have communications subsystems that communicate alarms or faults, store detection system data and allow operators to add explanatory information to alarm records. This section of the design defines the specifications for the communications system, taking into account data needs for the traffic volume expected; the desired amount of information and number of camera images per occupancy or alarm event; real time versus non-real time observation; the length of time that data are stored; the level and method of information protection; feature specifications, including integration needs with other internal or external systems; configuration specifications, including locations and types of communications links, such as workstations, alarm panels, phones and radios; and reporting specifications from the detection systems.

Handheld equipment

I–35. This section lists the type and quantity of handheld radiation detection equipment needed at the site.

Other equipment

I–36. This section defines any additional equipment needed at the secondary inspection areas, such as inspection platforms, booths, portable computers and safety equipment.

Interference factors

I–37. This section describes any features at the site that might interfere with the operation of the equipment. These features might include vehicle or pedestrian traffic congregating near the radiation portal monitor; lack of space for secondary inspections; or nearby equipment that emits radiation, such as X ray machines, industrial or medical sources, or vehicle and cargo inspection systems.

OTHER CONSIDERATIONS

Storage facilities on the site or within the State

I–38. This section lists any existing temporary radioactive material storage facilities that are currently operational on the site or within the State where intercepted material could be temporarily stored, and how material could be safely and securely transported there.

Training plan and technical know-how

I–39. The organizations responsible for radiation equipment operation, training and maintenance are defined in the concept of operations (see para. I–17). This section details the operation and maintenance training needs for front line officers to support the detection operation. Training courses might include the following:

(a) Basic radiation protection and safety training;
(b) Front line officer detection techniques training;
(c) Technical expert support training;
(d) Handheld acceptance testing and maintenance training;
(e) Equipment corrective and preventive maintenance training;
(f) Response and source recovery training.

I–40. A shift schedule and staffing plan for operations and maintenance is developed and informs the training needs. A training plan details the current training programmes, the organizational needs for training staff, the number of people that need to be trained, the organizations responsible for providing training and qualified instructors, and the frequency of training programmes based on staffing needs and rotation policies.

SUSTAINABILITY AND MAINTENANCE

I–41. This section describes the plan for the sustainability and maintenance of the detection operation. Plans for ongoing training, maintenance and the sustainability of resources and funding need to be included. The organization responsible for the maintenance of the equipment will need to decide whether it will maintain the equipment itself using its own technical experts or use contractors with the necessary training and experience to maintain and repair the equipment.

PROJECT RISKS, POTENTIAL ISSUES AND RISK MITIGATION STRATEGY

I–42. Any risks, opportunities or potential issues are noted in this section if not already described.

Annex II

EXAMPLES OF CONTENT FOR STANDARD
OPERATING PROCEDURES

II–1. Standard operating procedures describe the specific steps that front line officers take to carry out their duties related to detection systems and measures. The descriptions that follow provide examples of the type of information that could be included in the standard operating procedures at State borders.

II–2. Standard operating procedures for points of exit or entry (POEs) begin with an introductory section that covers the following information:

(a) Goals: description (often derived from a higher level strategy document) of the goals and subsequent operating procedures of the competent authority to control and monitor movement of radioactive material at the identified designated POE or border area. The section can include detection and related initial response procedures, as well as maintenance guidelines for deployed equipment and technology systems.
(b) Legislative basis: description of national legislation relating to standard operating procedures. Standard operating procedures derive from the concept of operations and complement it with respect to tactical operations at the designated POE or other border area.
(c) Authorization of personnel: description of operational training needed to allow the front line officers of the competent authority to implement radiation detection measures and procedures. Identification of the know-how to ensure personal safety makes front line officers confident in operating procedures.
(d) Personnel: listing of the competent authority's personnel at the designated POE or border area and their general functions (i.e. head of unit – management and organization; shift supervisor – operational supervision; operator – implementer of detection measures; inspector – assistant to operator).
(e) Detection process: overall description of the consecutive steps of detection measures, which largely derive from the national and site level concept of operations. This section can include a description of general detection and initial assessment steps, secondary inspection, technical support and initiation of response.

EXAMPLE OF CONTENT FOR STANDARD OPERATING PROCEDURES AT A DESIGNATED POINT OF EXIT OR ENTRY

General duties and responsibilities

II–3. This section of the standard operating procedures details the roles and responsibilities of all personnel and describes the steps they take for different situations.

II–4. Because personnel at designated POEs have multiple roles and responsibilities to manage, only one of which is related to detection systems and measures, standard operating procedures define personnel duties as they relate to such activities. Examples of personnel duties at designated POEs are provided below:

(a) Head of the unit at the designated POE: manages and organizes radiation detection processes; ensures quality control and the availability of human and technical resources; tracks training needs; plans logistics.

(b) Shift supervisor at the designated POE: gives an operational briefing to the shift personnel; receives information alerts and risk analysis results; designates inspection team members and coordinates their activities; supervises routine operations and detection processes; supports secondary inspections; supports response measures; is responsible for detection related reporting.

(c) Operator at the designated POE: monitors equipment health status and proper usage and communicates the results; performs basic hardware and software maintenance; operates the workstation; leads the inspection team; conducts the initial assessment; conducts the secondary inspection; confirms false or innocent alarms; identifies the need for response; defines a safe perimeter, when needed; cooperates with subject matter experts of technical expert support organizations; performs computer based processing; performs after-incident reporting.

(d) Inspector at the designated POE: operates the workstation and ensures proper use of detection equipment; is a member of the inspection team; conducts the initial assessment; conducts the secondary inspection; confirms false or innocent alarms; defines a safe perimeter, when needed.

(e)   Operator of the national communications system[1], if applicable: supervises the operations of subordinated and designated POEs at the regional or national level; ensures continuity of communication with subject matter experts, technical support, emergency services and other agencies involved in the response at the designated POE; performs statistical processing of alarms and reporting.

Standard operating procedures by alarm type

II–5. Standard operating procedures can be organized in different ways. The following example details steps according to alarm type:

(a)   False alarm: the operator at the designated POE types the appropriate reason — such as 'radiation background change' or 'inappropriate use of radiation portal monitor' — into the graphical user interface on the workstation and closes the alarm notification.

(b)   Innocent alarm:
   (i)   The operator at the designated POE conducts the initial assessment of the alarm.
   (ii)   After confirmation of the presence of ionizing radiation, the operator at the designated POE checks the documentation and performs the secondary inspection by instrument to identify the source of radiation and its legality (security and safety measures are considered).
   (iii) After confirmation of the legality of the shipment of radioactive material, the methodology described in Annex III is followed. People who are confirmed to be under radionuclide treatment may proceed with border crossing.
   (iv)   The operator at the designated POE types the appropriate reason — such as 'transport of naturally occurring radioactive material', 'legal shipment' or 'medical treatment' — into the graphical user interface on the workstation, with specific descriptions such as the person's identification or the cargo manifest, and closes the alarm notification.

(c)   Non-innocent alarm:
   (i)   The operator at the designated POE conducts the initial assessment of the alarm.

---

[1] The national communications system is an information system that unifies nationwide subordinated and designated POE detection systems and receives alarms from them online in near real time.

(ii) After confirmation of the presence of ionizing radiation, the operator at the designated POE checks the documentation and performs the secondary inspection by instrument to identify the source of radiation and its legality (security and safety measures are considered).

(iii) The operator at the designated POE confirms one or more of the following: the presence of radiation above the threshold defined by national legislation; violation of transportation and/or packaging rules; the possibility of illicit trafficking of nuclear and/or other radioactive material. The operator at the designated POE then informs the supporting response organization.

(iv) After finalization of response measures, the operator at the designated POE types the appropriate reason — such as 'violation of transport and/or packaging rules', 'orphan source in the shipment' or 'illicit trafficking' — into the graphical user interface on the workstation. The operator then uploads the appropriate supporting data from the radionuclide identification device to the computer and attaches them to the alarm record, enters a detailed description of detection and response measures with identification data, and closes the alarm.

Standard security and safety measures during the secondary inspection

II–6. The following security and safety measures need to be taken during the secondary inspection:

(a) If a source emits 0.1 mSv/h or more at a distance of 1 m (the radiation level might differ according to national legislation), the secondary inspection is stopped immediately (corresponding radiation levels shown on instruments are provided in the equipment manual and in training manuals) and an inner cordon is defined.

(b) A safe distance is marked with an outer cordon (indicating where the dose rate is 0.2 mSv/h or less), and the safe area is marked with warning signs (e.g. yellow tape).

(c) No one is allowed to enter the marked area until the response team completes its activities and clears the area.

(d) Personnel not participating in the inspection procedures are removed from the area.

(e) If possible, the suspect vehicle, suspect cargo or suspect individual is taken to a safe area.

(f) Vehicles, cargo or luggage that contain nuclear or other radioactive material are secured and isolated by a specialized safety barrier. The cargo owners and other individuals who might have had contact with the cargo are moved to an isolated room for inspection.

(g) After the source has been secured, sealed and transported away by the appropriate entities, and after the area has been decontaminated, the shift supervisor at the designated POE makes the decision to continue border protection activities.

EXAMPLE OF CONTENT FOR STANDARD OPERATING PROCEDURES IN A BORDER AREA

General duties and responsibilities

II–7. Similar to the standard operating procedures for designated POEs, this section of the standard operating procedures for border areas details the roles and responsibilities of all personnel, and then describes the steps to be taken in different situations.

II–8. Control of the movement of nuclear or other radioactive material is one of several types of border protection activity carried out, in addition to surveillance, patrolling and border checks. Radiation detection measures are incorporated into existing border security activities. Examples of responsibilities of border unit personnel are described below:

(a) Head of the border unit: manages and organizes overall border security operations, including radiation detection; ensures quality control and availability of human and technical resources; tracks training needs; plans logistics.

(b) Operational commander or shift supervisor: gives an operational briefing to the shift personnel; receives information alerts and risk analysis results; supervises routine operations including detection measures; is responsible for detection related reporting.

(c) Surveillance operator or operational supervisor: operates border surveillance systems; verifies the health status of detection instruments and communicates the results; coordinates and communicates with front line officers and field patrols; assists with information verification. This operator could be trained as a specialist in conducting radionuclide identification.

(d) Border patrol: operates detection equipment in the field; conducts the initial assessment; conducts the secondary inspection; confirms false or innocent alarms; defines a safe perimeter, when needed.

Standard operating procedures by alarm type

II–9. Standard operating procedures can be organized in different ways. The following example details steps according to alarm type for a border area:

(a) False alarm: the border patrol conducts the initial assessment of the primary detection, confirms there is no radiation, and continues normal border security operations.

(b) Innocent alarm:
   (i) The border patrol conducts the initial assessment of the primary detection, gathers information and confirms the presence of ionizing radiation.
   (ii) After confirmation of the presence of ionizing radiation, the border patrol checks the documentation and performs the secondary inspection by instrument to identify the source of radiation and its legality (security and safety measures are considered).
   (iii) People who are confirmed to be under radionuclide treatment, or nuclear or other radioactive material that is confirmed to be legal (e.g. material with radiation levels lower than legal thresholds and that has been confirmed not to be a threat) may be cleared to proceed with border crossing, but other border security operations are completed before a decision is made, in accordance with law enforcement procedures.
   (iv) The surveillance operator or operational supervisor records the relevant information related to the detection.

(c) Non-innocent alarm:
   (i) The border patrol conducts the initial assessment of the primary detection, gathers information and confirms the presence of ionizing radiation.
   (ii) After confirmation of the presence of ionizing radiation, the border patrol checks the documentation and performs the secondary inspection by instrument to identify the source of radiation and its legality (security and safety measures are considered).
   (iii) The border patrol confirms one or more of the following: the presence of radiation above the threshold defined by national legislation; violation of transportation and/or packaging rules; the possibility of illicit trafficking of nuclear and/or other radioactive material. The border patrol then informs the supporting response organization.
   (iv) After finalization of response measures, the surveillance operator or operational supervisor records the relevant information related to the detection.

Standard security and safety measures during the secondary inspection

II–10. The following security and safety measures need to be taken during the secondary inspection:

(a)   If a source emits 0.1 mSv/h or more at a distance of 1 m (the radiation level might differ according to national legislation), the secondary inspection is stopped immediatcly (corresponding radiation levels shown on instruments are provided in the equipment manual and in training manuals) and an inner cordon is defined.

(b)   A safe distance is marked with an outer cordon (indicating where the dose rate is 0.2 mSv/h or less), and the safe area is marked with warning signs (e.g. yellow tape).

(c)   No one is allowed to enter the marked area until the response team completes its activities and clears the area.

(d)   Personnel not participating in the inspection procedures are removed from the area.

(e)   If possible, the suspect vehicle, suspect cargo or suspect individual is taken to a safe area.

(f)   Vehicles, cargo or luggage that contain nuclear or other radioactive material are secured and isolated by a specialized safety barrier. The cargo owners and other individuals who might have had contact with the cargo are moved to an isolated room for inspection.

(g)   After the source has been secured, sealed and transported away by the appropriate entities, and after the area is cleared and safe, the operational commander makes the decision to continue border protection activities.

Annex III

EXAMPLE OF AN EVALUATION PROCESS FOR
ALARMS ON DECLARED SHIPMENTS

III–1. Although the movement of nuclear and other radioactive material is highly regulated and well controlled, illicit trafficking of such material even within a legal shipment might still occur.

III–2. Technological measures to detect and respond to the illicit trafficking of nuclear and other radioactive material have been and continue to be developed. As the verification of quantities and types of material crossing borders contributes to domestic and international control over such material, Member States need both procedural and technological measures to detect the movement of illicit material that might be concealed in otherwise legal shipments.

III–3. The objective of this annex is to provide national authorities, particularly customs officers, with a risk informed methodology to detect illicit trafficking of nuclear and other radioactive material within all declared shipments, including declared shipments of nuclear or other radioactive material. This is part of a larger process to identify any high risk shipment through the customs transaction process.

III–4. The World Customs Organization (WCO) has defined a framework of standards [III–1] that serve as the basic guidelines in the arrangement of a State's activities by its customs administration. One of the roles that customs officials perform is the checking of qualitative and quantitative aspects of goods being transferred across the border for consistency with the information on declared shipments.

SCENARIOS FOR ILLICIT TRAFFICKING OF NUCLEAR AND OTHER RADIOACTIVE MATERIAL

III–5. All shipments of radioactive material are required to comply with the Transport Regulations, the current edition of which is IAEA Safety Standards Series No. SSR-6 (Rev. 1), Regulations for the Safe Transport of Radioactive Material, 2018 Edition [III–2]. Compliance with SSR-6 (Rev. 1) [III–2] includes classification of the material to be transported, selection of the appropriate package type and preparation of the package for transport, including marking, labelling and shipping documentation.

III–6. Although the transport of nuclear and other radioactive material is carefully regulated for safety and security purposes, there are several possible scenarios for illicit trafficking of nuclear or other radioactive material out of regulatory control within a shipment. Examples of such scenarios are the following:

(a) Within the package: substitution, addition or removal of declared nuclear or other radioactive material.
(b) Within the conveyance or container:
— Substitution of packages containing nuclear or other radioactive material;
— Addition or removal of packages containing nuclear or other radioactive material;
— Replacement of packages with empty radioactive material packages (e.g. scams);
— Replacement of packages with other radioactive material packages that have different radioactive contents (with the same external packaging).

In addition, documentation might be falsified to disguise illicit trafficking of nuclear and other radioactive material as a legal shipment.

PROCESS TO DETECT ILLICIT TRAFFICKING OF NUCLEAR AND OTHER RADIOACTIVE MATERIAL

III–7. Figure III–1 provides a high level outline of a graded approach to determine whether a declared shipment contains any illicit nuclear or other radioactive material.

Process for selecting declared shipments for inspection

Threat assessment

III–8. If the possibility of a malicious act using nuclear or other radioactive material is identified as a threat at the national level, then illicit cross-border trafficking of nuclear or other radioactive material using declared shipments as a mode of transport will be a risk that needs to be managed by the front line officer and other competent authorities.

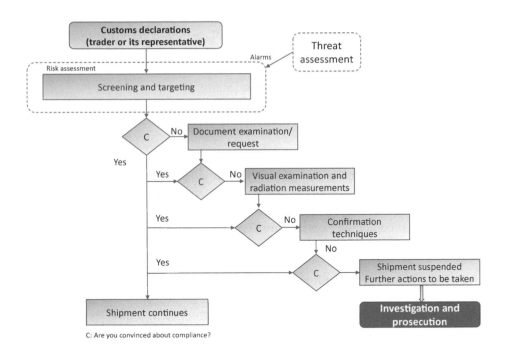

FIG. III–1. A high level outline of a graded approach to detect illicit trafficking of nuclear and other radioactive material (courtesy of N. Kravchenko, National Research Nuclear University, Russian Federation).

Risk assessment

III–9. Through the risk assessment process, the competent authority at the border makes targeted decisions about the allocation of control resources at the operational level.

III–10. For the identification of risks, the gathering and processing of information on potential risks is needed. For customs administrations, the WCO has issued several guidance documents on risk assessment, as well as the following framework of standards:

(a) Revised Kyoto Convention (2008) [III–3];
(b) SAFE Framework of Standards [III–1];
(c) WCO Customs Risk Management Compendium, Vols 1 and 2 [III–4].

III–11. Sources that can be used to gather information concerning nuclear and other radioactive material include the following:

(a) The IAEA's Incident and Trafficking Database;
(b) Customs seizure and other law enforcement databases;
(c) Customs declarations and other historical data;
(d) Intelligence.

III–12. If the consignee is an authorized user of nuclear or other radioactive material, the shipper is recognized as reputable, and if all other risk assessment elements lead to identifying the consignment as compliant and therefore of low risk, then these combined elements might provide sufficient information to allow the goods to be moved across the border unhindered by customs officers.

Screening and targeting shipments

III–13. Customs declarations, manifest information, bills of lading, authorizations and any other documentation that might be available for front line officers contain important information against which risk indicators and profiles are assessed to determine the level of risk associated with an individual shipment.

III–14. By using indicators, profiles and experience, trained front line officers are able to evaluate trade patterns and anomalies that might imply an elevated risk of illicit trafficking. This may result in targeting such shipments or consignments for further investigative action.

III–15. Competent authorities need to develop criteria to guide their screening and targeting practices. Screening and targeting criteria are sensitive information and need to be treated as such.

III–16. Customs administrations might operate automated systems incorporating risk indicators and profiles. If the front line officer is not from a customs administration, the front line officer needs to be able to access the information available in the system. Automation facilitates the screening process, although risk indicators and profiles can also be implemented without automation.

Control process

III–17. Depending on the outcome of the process for selecting declared shipments for inspection, goods may either be released or subjected to additional controls. As outlined in Fig. III–1, customs control techniques can be categorized into three general types:

(a) Document examination;
(b) Visual examination and radiation measurements;
(c) Confirmation techniques.

Document examination

III–18. Document examination is often the first stage of the control process if additional measures to verify compliance of the shipment are needed. The examination concentrates on reviewing customs declarations and transport related documentation. The following documentation might be needed to complete the clearance process:

(a) Shipping documents;
(b) Export, import, transit or transhipment permits;
(c) Dangerous goods declaration (when applicable).

III–19. The following information needs to be present in the documentation to make a valid assessment:

(a) Name, nature (physical and/or chemical form) and quantitative characteristics, such as activity or mass of the material;
(b) United Nations (UN) number, together with the proper shipping name;
(c) Legal name and address of the recipient (when applicable);
(d) Shipping and receiving States;
(e) Legal name and address of the shipper (when applicable).

III–20. In addition to the usual documentation that relates to the shipment, the front line officer can request other information from the shipper, such as the following:

(a) Dose rate monitoring records, if available, which can contain information such as gamma and neutron dose rates in contact with and at a specified distance from the package, and contact temperature;
(b) Certificate or permission for the handling of nuclear and other radioactive material, showing the date of issue and the expiration date;
(c) When applicable, the certificate or permission for the safe transport or conveyance of nuclear and other radioactive material, showing the date of issue and the expiration date;

(d) Certificate for the packaging used to carry the radioactive material, as described in SSR-6 (Rev. 1) [III–2].

III–21. Information extracted from the documents can be analysed to detect inconsistencies, anomalies or illogical combinations. For example, the transport index (TI) of a package containing a pure beta emitter of small activity cannot be high. Significant discrepancies can trigger further actions.

Visual examination and radiation measurements

III–22. Visual examination often consists of the checking of seals. The radiation measurements described in this section might not prove the legitimacy of a shipment to a high degree of confidence. These checks have to be used in conjunction with other available documents and data as part of an overall assessment. The radiation measurements can be initiated by measuring the dose rate and confirming the TI, as described in SSR-6 (Rev. 1) [III–2].

III–23. Where available and appropriate, dose rate measurements could be made for broad comparison with the transport documents. The dose rates of shipments of short lived nuclides (with a half-life typically less than five days) will be lower than the values given in the transport documents. The dose rate measurements made at a border crossing might be different than those made at the point of origin owing to the limitations in the accuracy of radiation monitors. In addition, in the case of a multipackage shipment, the dose rate measurement of one package can be affected by radiation from other packages. These discrepancies need to be taken into account while making dose rate measurements. Radioactive material in excepted packages does not require labelling as radioactive but will be marked with the appropriate UN number. Dose rates at external surfaces of excepted packages could be up to 5 μSv/h. When measured, the calculated TI might not match the transport documents and the labels exactly owing to a number of factors, including the following:

(a) Different points of measurement;
(b) Different devices used for measurements;
(c) Different calibration criteria;
(d) Different environmental conditions (e.g. radiation background, proximity to other radioactive material packages, humidity, temperature).

III–24. All radiation measurement instruments (e.g. handheld, portable, installed monitors) need to be used and maintained in accordance with the equipment manuals and need to be calibrated. Records of maintenance and calibration need to be maintained in accordance with the national requirements.

III–25. The labels affixed to the external surfaces of packages, overpacks and freight containers indicate the TI. The TI is related to the dose rate measured on the exterior of the packages, overpacks and freight containers.

III–26. Before starting any measurements on a shipment, appropriate radiation protection procedures have to be implemented and followed (in accordance with the requirements for radiation protection as established in IAEA Safety Standards Series No. GSR Part 3, Radiation Protection and Safety of Radiation Sources: International Basic Safety Standards [III–5]). The measurements described in paras III–22 to III–34 involve trained and competent personnel, procedures, and appropriate equipment to measure the dose rate on the exterior of a package without the need to open the package.

III–27. SSR-6 (Rev. 1) [III–2] states:

"523. The *TI* for a *package*, *overpack* or *freight container*, or for unpackaged *LSA-I*, *SCO-I* or *SCO-III*, shall be the number derived in accordance with the following procedure:

(a) Determine the maximum *dose rate* in units of millisieverts per hour (mSv/h) at a distance of 1 m from the external surfaces of the *package*, *overpack*, *freight container* or unpackaged *LSA-I*, *SCO-I* and *SCO-III*. The value determined shall be multiplied by 100. For *uranium* and thorium ores and their concentrates, the maximum *dose rate* at any point 1 m from the external surface of the load may be taken as:
(i) 0.4 mSv/h for ores and physical concentrates of *uranium* and thorium;
(ii) 0.3 mSv/h for chemical concentrates of thorium;
(iii) 0.02 mSv/h for chemical concentrates of *uranium*, other than uranium hexafluoride.
(b) For *tanks*, *freight containers* and unpackaged *LSA-I*, *SCO-I* and *SCO-III*, the value determined in step (a) shall be multiplied by the appropriate factor from [Table III–1].
(c) The value obtained in steps (a) and (b) shall be rounded up to the first decimal place (for example, 1.13 becomes 1.2), except that a value of 0.05 or less may be considered as zero and the resulting number is the *TI* value.

"524. The *TI* for each rigid *overpack, freight container* or *conveyance* shall be determined as the sum of the *TI*s of all the *packages* contained therein. For a *shipment* from a single *consignor*, the *consignor* may determine the *TI* by direct measurement of *dose rate*.

"524A. The *TI* for a non-rigid *overpack* shall be determined only as the sum of the *TI*s of all the *packages* within the *overpack.* "[1]

III–28. The following is an example of how to determine the TI for a package. Figure III–2 shows the package and the highest dose rates measured at a distance of 1 m from its external surface. Therefore, the maximum radiation level at a distance of 1 m from the external surface of the package is 0.02 mSv/h. The TI of the package is calculated by multiplying the maximum radiation level at a distance of 1 m from the external surface (i.e. 0.02) by 100, and the result is 2.0.

---

[1] SSR-6 (Rev. 1) [III–2] provides the following definitions for low specific activity (LSA) and surface contaminated object (SCO):

In para. 409(a), LSA-I is defined as "(i) *Uranium* and thorium ores and concentrates of such ores, and other ores containing naturally occurring radionuclides. (ii) *Natural uranium, depleted uranium*, natural thorium or their compounds or mixtures, that are unirradiated and in solid or liquid form. (iii) *Radioactive material* for which the A_2 value is unlimited. *Fissile material* may be included only if excepted under para. 417. (iv) Other *radioactive material* in which the activity is distributed throughout and the estimated average *specific activity* does not exceed 30 times the values for the activity concentration specified in paras 402–407. *Fissile material* may be included only if excepted under para. 417."

In para. 413(a), SCO-I is defined as "A solid object on which: (i) The *non-fixed contamination* on the accessible surface averaged over 300 cm^2 (or the area of the surface if less than 300 cm^2) does not exceed 4 Bq/cm^2 for beta and gamma emitters and *low toxicity alpha emitters*, or 0.4 Bq/cm^2 for all other alpha emitters; (ii) The *fixed contamination* on the accessible surface averaged over 300 cm^2 (or the area of the surface if less than 300 cm^2) does not exceed 4 × 10^4 Bq/cm^2 for beta and gamma emitters and *low toxicity alpha emitters*, or 4000 Bq/cm^2 for all other alpha emitters; (iii) The *non-fixed contamination* plus the *fixed contamination* on the inaccessible surface averaged over 300 cm^2 (or the area of the surface if less than 300 cm^2) does not exceed 4 × 10^4 Bq/cm^2 for beta and gamma emitters and low toxicity alpha emitters, or 4000 Bq/cm^2 for all other alpha emitters."

In para. 413(c), SCO-III is defined as "A large solid object which, because of its size, cannot be transported in a type of package described in these Regulations and for which: (i) All openings are sealed to prevent release of *radioactive material* during conditions defined in para. 520(e); (ii) The inside of the object is as dry as practicable; (iii) The *non-fixed contamination* on the external surfaces does not exceed the limits specified in para. 508; (iv) The *non-fixed contamination* plus the *fixed contamination* on the inaccessible surface averaged over 300 cm^2 does not exceed 8 × 10^5 Bq/cm^2 for beta and gamma emitters and *low toxicity alpha emitters*, or 8 × 10^4 Bq/cm^2 for all other alpha emitters."

TABLE III–1. MULTIPLICATION FACTORS FOR TANKS, FREIGHT CONTAINERS AND UNPACKAGED LSA-I, SCO-I AND SCO-III

Size of load[a]	Multiplication factor
size of load $\leq$ 1 m^2	1
1 m^2 < size of load $\leq$ 5 m^2	2
5 m^2 < size of load $\leq$ 20 m^2	3
20 m^2 < size of load	10

Note: This table is a reproduction of table 7 in SSR-6 (Rev. 1) [III–2].
[a] Largest cross-sectional area of the load being measured.

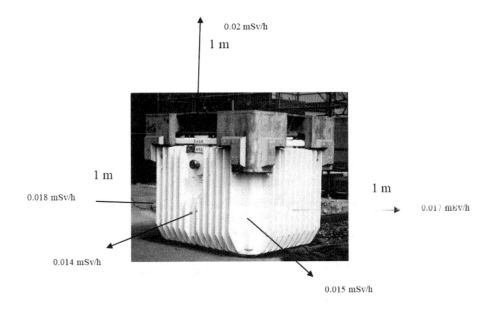

FIG. III–2. Radiation source with radiation measurements at 1 m (courtesy of N. Kravchenko, National Research Nuclear University, Russian Federation).

III–29. The customs officers need to be cautioned that in certain instances the measured value of the TI of a consignment could differ from the declared value. The following are examples of such instances:

(a) Example 1: In the case of radionuclides used in nuclear medicine and short lived nuclides, the measured and declared TI values could be significantly different.

(b) Example 2: When packages are shipped in an overpack or freight container, the TI calculations are for the overpack or freight container, not the individual packages. Comparison of the TIs in such instances would be erroneous and problematic.

III–30. Paragraph 529 of SSR-6 (Rev. 1) [III–2] states:

"*Packages, overpacks* and *freight containers* shall be assigned to either category I-WHITE, II-YELLOW or III-YELLOW in accordance with the conditions specified in [Table III–2] and with the following requirements:

(a) For a *package, overpack* or *freight container*, the *TI* and the surface *dose rate* conditions shall be taken into account in determining which category is appropriate. Where the *TI* satisfies the condition for one category but the surface *dose rate* satisfies the condition for a different category, the *package, overpack* or *freight container* shall be assigned to the higher category. For this purpose, category I-WHITE shall be regarded as the lowest category.

(b) The *TI* shall be determined following the procedures specified in paras 523, 524 and 524A.

(c) If the surface *dose rate* is greater than 2 mSv/h, the *package* or *overpack* shall be transported under *exclusive use* and under the provisions of paras 573(a), 575 or 579, as appropriate.

(d) A *package* transported under a *special arrangement* shall be assigned to category III-YELLOW except under the provisions of para. 530.

(e) An *overpack* or *freight container* that contains *packages* transported under *special arrangement* shall be assigned to category III-YELLOW except under the provisions of para. 530."

III–31. The typical parameters of the equipment used for confirmation of the TI, category and radiation dose rate of packages are shown in Table III–3.

III–32. Portable dose rate meters can be used to confirm the package category and the TI. The customs officers need to be aware before reaching any judgement that when using portable dose rate meters of the types in Table III–3, there could be relatively large differences between the measurements of different dose rate meters.

TABLE III–2. CATEGORIES OF PACKAGES, OVERPACKS AND FREIGHT CONTAINERS

Conditions		Category
TI	Maximum *dose rate* at any point on external surface	
0[a]	Not more than 0.005 mSv/h	I-WHITE
More than 0 but not more than 1[a]	More than 0.005 mSv/h but not more than 0.5 mSv/h	II-YELLOW
More than 1 but not more than 10	More than 0.5 mSv/h but not more than 2 mSv/h	III-YELLOW
More than 10	More than 2 mSv/h but not more than 10 mSv/h	III-YELLOW[b]

Note: This table is a reproduction of table 8 in SSR-6 (Rev. 1) [III–2].
[a] If the measured *TI* is not greater than 0.05, the value quoted may be zero in accordance with para. 523(c) [of SSR-6 (Rev. 1) [III–2]].
[b] Shall also be transported under *exclusive use* except for *freight containers* (see table 10 [in SSR-6 (Rev. 1) [III–2]]).

TABLE III–3. TYPICAL PERFORMANCE CHARACTERISTICS FOR PORTABLE DOSE RATE METERS

Type of portable dose rate meter	Measured parameter	Measured range	Energy range of the measured radiation	Maximum error (%)
Gamma	Dose rate (μSv/h)	From 0.1 to 1×10^4 μSv/h	From 0.05 to 3 MeV	±20
Neutron	Dose rate (μSv/h)	From 1.0 to 1×10^4 μSv/h	From thermal to 14 MeV	±40

III–33. Operating specifications of typical handheld radiation detection instruments include the following:

(a) Working temperature: from –20°C to +50°C.
(b) Duration of continuous work using the built-in batteries: not less than eight hours.
(c) Weight of handheld instruments: typically less than 5 kg.
(d) Total measuring time: not more than 300 seconds (typically, measurement times could vary from 10 to 100 seconds).

III–34. Appropriate radiation protection procedures have to be implemented and followed (in accordance with the requirements for radiation protection as established in GSR Part 3 [III–5]).

Confirmation techniques

III–35. It might be necessary to determine the contents of a package with greater confidence. Confirmation of the package contents can be achieved by qualitative and quantitative measurements of the declared shipment of nuclear and other radioactive material.

III–36. The appropriate measurement techniques need to be selected to allow for confirmation of the declared shipment. The capability to conduct these measurements might not always exist within the front line officer's organization. Front line officers need to cooperate with the relevant regulatory bodies or expert organizations, as necessary.

III–37. If the shipment needs to be held for further investigation, the competent authority needs to identify an appropriate secure storage area where the consignment can be stored in conformity with applicable regulations and to provide appropriate security arrangements until the investigation has been completed and the consignment can be moved.

III–38. It needs to be ensured that confirmation techniques do not damage or change the integrity of the package or the characteristics of the verified nuclear material or radioactive material. The confirmation techniques described in paras III–39 to III–54 involve trained and competent personnel, procedures, and appropriate equipment to confirm the contents of a package without the need to open the package.

III–39. The following confirmation techniques can be used, where applicable, separately or in combination, to assess the declared contents of a shipment of nuclear or other radioactive material:

(a) Weighing of a package;
(b) X ray examination;
(c) Neutron measurements;
(d) Gamma spectrometry.

Weighing of a package

III–40. Paragraph 533 of SSR-6 (Rev. 1) [III–2] states that: "Each *package* of gross mass exceeding 50 kg shall have its permissible gross mass legibly and durably marked on the outside of the *packaging*". In many cases where the gross mass is 50 kg or less, the gross mass of the package can be stated in the transport documents. Weighing packages and comparing the measured mass with the documented mass is one of the methods of assessing the contents or the shielding of a package, or both. For instance, a significant discrepancy in the declared mass and actual mass of the package might indicate that additional or strengthened shielding has been added, which could be an indicator that nuclear or radioactive material has been added or substituted.

III–41. The weighing equipment needs to be calibrated and maintained in accordance with the equipment manual. Records of maintenance and calibration need to be maintained in accordance with the national requirements. Limitations on the accuracy of weight measurements at the point of origin and standard errors in the user measurements need to be considered when addressing weight discrepancies.

III–42. The weighing of packages provides only limited information since the declared mass of the package is its maximum weight. The stillage playing the function of a restraint system attached to the conveyance, as applicable, might not be part of the package and therefore would not be included in the declared mass. It needs to be ensured that the stillage is not removed from the package during transport.

X ray examination

III–43. X ray or gamma ray imaging of packages can be useful to confirm the contents and geometry of packages containing nuclear or other radioactive material. Standard and special X ray equipment (radiography equipment) can be used for this purpose. In some cases, the density of the package material, the contents of the package or the presence of shielding material will make interpretation of a radiography image difficult or impossible. X ray examination can be more useful for shipments not declared as shipments of nuclear or radioactive material, where the presence of shielding material might be an indicator of attempted concealment of nuclear or other radioactive material.

Neutron measurements

III–44. Caution is needed in the selection of the instrument used to measure neutrons, since the absence of neutron counts might not confirm the absence of nuclear material.

III–45. Radiation portal monitors with neutron detection capabilities provide a reliable way of detecting the presence of a neutron source. Radiation portal monitors can indicate the presence of illicit material if the declared nuclear or other radioactive material is not a known neutron emitter.

III–46. Neutron dose rate meters and handheld neutron survey meters can be used as measurement tools for high level neutron dose rates. However, users need to be aware of the sensitivity limitations of the equipment and interpret the readings with caution; in particular, results showing zero counts might not be correct.

Gamma spectrometry

III–47. Gamma spectrometry can be used to establish the type and quantity of gamma emitting radionuclides, including the enrichment of nuclear material. Qualitative measurements can be made by a front line officer; this demands limited training. If a front line officer's organization wants to conduct quantitative measurements, the steps in paras III–48 to III–50 need to be considered.

III–48. An appropriate set-up has to be established for conducting the measurements. Suitably qualified and experienced personnel need to be entrusted with the task of conducting the measurements. It is desirable to have a dedicated measurement area away from other radioactive material to reduce background radiation and to improve measurement accuracy. In addition, collimators can be used for the detectors. Knowledge of the package design is essential for qualitative and quantitative measurements of radioactive material. Pertinent information (including drawings) can be obtained from the regulatory body, if requested by the customs officers.

III–49. It is advised to check the package design to locate the radiation source position inside the package. Requirements for the disclosure of design characteristics of transport packages are described in para. 838(j) of SSR-6 (Rev. 1) [III–2]. If the data are not available, it is necessary to verify the radiation source position by making additional measurements around the package. The maximum count rate or dose meter reading will correspond to the actual position of the radioactive sources in the package, as shown in Fig. III–3. The distance from the package surface needs to be selected so that the count rate is optimized. The spectrometer dead time needs to be below the value indicated in the equipment manual.

III–50. The measured gamma spectrum is analysed to identify the radionuclides and calculate their activities. Information on the declared radionuclides, their activities on a specified date, the type of package, and the distance from the detector to the surface of the package can be used to confirm the declared contents. Software that incorporates package design characteristics might be available to assist in the analysis. Typical parameters of a gamma spectrometer are shown in Table III–4.

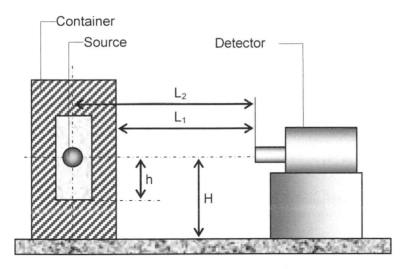

FIG. III–3.   Position of detector and package (courtesy of N. Kravchenko, National Research Nuclear University, Russian Federation).
L_1 — distance of the detector from the external surface of the container;
L_2 — distance of the detector from the source position within the container;
H — distance of the source position within the container from the floor;
h — distance of the source position from the bottom of the containment system within the container.

TABLE III–4. TYPICAL PARAMETERS OF A GAMMA SPECTROMETER

Parameter	Value
Range of measured gamma energies (keV)	$50 - 3 \times 10^3$
Energy resolution:	
— For semiconductor spectrometer	>0.2%
— For scintillation spectrometer	>8%
Efficiency (1332 keV, ^{60}Co):	
— For semiconductor spectrometer	>15%
— For scintillation spectrometer	>40%
Maximum error in quantitative measurement (point geometry)	±10%
Continuous measurement time:	
— Main power supply	min. 24 h
— Batteries	min. 8 h
Number of channels in the analyser:	
— Semiconductor spectrometer	8192
— Scintillation spectrometer	1024
Environmental conditions for stable spectrometer operation:	
— Temperature (°C)	−20 to +50
— Relative humidity (%)	≤90

Confirmation of uranium isotopic composition

III–51. If there is a need to confirm a uranium shipment, this confirmation has to include the confirmation of the enrichment.

III–52. Computer software is available to assist in the confirmation of the declared uranium enrichment, which uses different energy regions of the collected gamma spectra of uranium or a combination of two of them: 89–100 keV, 185 keV and 1001 keV. Owing to high attenuation of low energy gamma rays, the 89–100 keV region is typically used for packages with a steel wall thickness of less than 5 mm. Only authorized software is to be used for this purpose.

III–53. If similar packages containing the same or similar uranium products are found in several shipments, one of the packages with known uranium enrichment can be selected as a reference. The other packages can be measured in the same geometry used for the reference package, and the results can be compared.

Confirmation of plutonium isotopic composition

III–54. Computer software can be used to assist in the analysis of plutonium isotopic composition. Plutonium isotopic composition can be measured using three regions of the gamma spectrum: 94–105 keV, 120–460 keV or 630–770 keV. The low energy region is preferable, but it can be used only for packages with low attenuation of gamma rays. Only authorized software is to be used for this purpose.

Evaluation results

III–55. If the shipment is suspended as a result of the selection and control process, the front line officer needs to notify the competent authority and await further instructions.

REFERENCES TO ANNEX III

[III–1] WORLD CUSTOMS ORGANIZATION, SAFE Framework of Standards, 2021 edn, WCO, Brussels (2021).
[III–2] INTERNATIONAL ATOMIC ENERGY AGENCY, Regulations for the Safe Transport of Radioactive Material, 2018 Edition, IAEA Safety Standards Series No. SSR-6 (Rev. 1), IAEA, Vienna (2018).
[III–3] CUSTOMS CO-OPERATION COUNCIL, International Convention on the Simplification and Harmonization of Customs Procedures (as amended), World Customs Organization, Brussels (2008).
[III–4] WORLD CUSTOMS ORGANIZATION, WCO Customs Risk Management Compendium, Vols 1 and 2, WCO, Brussels (undated).
[III–5] EUROPEAN COMMISSION, FOOD AND AGRICULTURE ORGANIZATION OF THE UNITED NATIONS, INTERNATIONAL ATOMIC ENERGY AGENCY, INTERNATIONAL LABOUR ORGANIZATION, OECD NUCLEAR ENERGY AGENCY, PAN AMERICAN HEALTH ORGANIZATION, UNITED NATIONS ENVIRONMENT PROGRAMME, WORLD HEALTH ORGANIZATION, Radiation Protection and Safety of Radiation Sources: International Basic Safety Standards, IAEA Safety Standards Series No. GSR Part 3, IAEA, Vienna (2014).

ORDERING LOCALLY

IAEA priced publications may be purchased from the sources listed below or from major local booksellers.

Orders for unpriced publications should be made directly to the IAEA. The contact details are given at the end of this list.

NORTH AMERICA

Bernan / Rowman & Littlefield
15250 NBN Way, Blue Ridge Summit, PA 17214, USA
Telephone: +1 800 462 6420 • Fax: +1 800 338 4550
Email: orders@rowman.com • Web site: www.rowman.com/bernan

REST OF WORLD

Please contact your preferred local supplier, or our lead distributor:

Eurospan Group
Gray's Inn House
127 Clerkenwell Road
London EC1R 5DB
United Kingdom

Trade orders and enquiries:
Telephone: +44 (0)176 760 4972 • Fax: +44 (0)176 760 1640
Email: eurospan@turpin-distribution.com

Individual orders:
www.eurospanbookstore.com/iaea

For further information:
Telephone: +44 (0)207 240 0856 • Fax: +44 (0)207 379 0609
Email: info@eurospangroup.com • Web site: www.eurospangroup.com

Orders for both priced and unpriced publications may be addressed directly to:

Marketing and Sales Unit
International Atomic Energy Agency
Vienna International Centre, PO Box 100, 1400 Vienna, Austria
Telephone: +43 1 2600 22529 or 22530 • Fax: +43 1 26007 22529
Email: sales.publications@iaea.org • Web site: www.iaea.org/publications